Approximability of Optimization Problems through Adiabatic Quantum Computation

Synthesis Lectures on Quantum Computing

Editor
Marco Lanzagorta, *U.S. Naval Research Labs*
Jeffrey Uhlmann, *University of Missouri-Columbia*

Quantum Walks for Computer Scientists
Salvador Elías Venegas-Andraca
2008

Approximability of Optimization Problems through Adiabatic Quantum Computation

William Cruz-Santos and Guillermo Morales-Luna

ISBN: 978-3-031-01391-1   paperback
ISBN: 978-3-031-02519-8   ebook

DOI 10.1007/978-3-031-02519-8

A Publication in the Springer series
SYNTHESIS LECTURES ON QUANTUM COMPUTING

Lecture #9
Series Editors: Marco Lanzagorta, *U.S. Naval Research Labs*
                Jeffrey Uhlmann, *University of Missouri-Columbia*
Series ISSN
Print 1945-9726   Electronic 1945-9734

Approximability of Optimization Problems through Adiabatic Quantum Computation

William Cruz-Santos
CU UAEM Valle de Chalco
Universidad Autónoma del Estado de México, México

Guillermo Morales-Luna
Computer Science Department, Cinvestav-IPN
Mexico City, Mexico

SYNTHESIS LECTURES ON QUANTUM COMPUTING #9

ABSTRACT

The *adiabatic quantum computation* (AQC) is based on the *adiabatic theorem* to approximate solutions of the Schrödinger equation. The design of an AQC algorithm involves the construction of a Hamiltonian that describes the behavior of the quantum system. This Hamiltonian is expressed as a linear interpolation of an *initial Hamiltonian* whose ground state is easy to compute, and a *final Hamiltonian* whose ground state corresponds to the solution of a given combinatorial optimization problem. The adiabatic theorem asserts that if the time evolution of a quantum system described by a Hamiltonian is large enough, then the system remains close to its ground state. An AQC algorithm uses the adiabatic theorem to approximate the ground state of the final Hamiltonian that corresponds to the solution of the given optimization problem.

In this book, we investigate the computational simulation of AQC algorithms applied to the MAX-SAT problem. A symbolic analysis of the AQC solution is given in order to understand the involved computational complexity of AQC algorithms. This approach can be extended to other combinatorial optimization problems and can be used for the classical simulation of an AQC algorithm where a Hamiltonian problem is constructed. This construction requires the computation of a sparse matrix of dimension $2^n \times 2^n$, by means of tensor products, where n is the dimension of the quantum system. Also, a general scheme to design AQC algorithms is proposed, based on a natural correspondence between optimization Boolean variables and quantum bits. Combinatorial graph problems are in correspondence with pseudo-Boolean maps that are reduced in polynomial time to quadratic maps. Finally, the relation among NP-hard problems is investigated, as well as its logical representability, and is applied to the design of AQC algorithms. It is shown that every monadic second-order logic (MSOL) expression has associated pseudo-Boolean maps that can be obtained by expanding the given expression, and also can be reduced to quadratic forms.

KEYWORDS

quantum computing, quantum algorithms, quantum information, adiabatic quantum computing, computer science, combinatorial optimization, treewidth, monadic second order logic

To my family and beloved Mari

WC

To my family

GM

Contents

Preface

This book is intended to give an introduction to adiabatic quantum computing from a computational point of view and it is our hope that the reader finds motivation in this fascinating field of research. During the writing of this book, the authors emphasized the importance of the mathematical and symbolic treatment of quantum computing, given the fundamentals of computational complexity, throughout the definition of the complexity classes, and the fundamental postulates of quantum mechanics. All of these concepts were unified using a homogeneous notation that enables us to state important computational problems in the language of quantum computing. The reader can find detailed developments of all stated problems and technical background, giving him or her the opportunity to make further contributions to the covered and related topics. An important aspect of this work is in the precise and detailed description of the Hamiltonian construction for adiabatic computation; this is important for classical simulation of the adiabatic evolution and the formulation of NP-Hard problem in the Hamiltonian language. Another important aspect of this book is in the logical representability of NP-hard problems. This proposal links the well-known results in descriptive logic with the Hamiltonian construction and design of adiabatic algorithms. This approach is in its infancy and the authors believe that it can be useful as a tool in the development and analysis of new quantum algorithms.

The organization of this book is as follows. In Chapter 2, a succinct introduction to computational complexity is given and the basic definitions of complexity classes and optimization problems are introduced. In Chapter 3, we give an introduction to adiabatic quantum computing. It is intended to be self-contained. In Chapter 4, we explore the classical simulation of the AQC for the MAX-SAT problem, and we propose a symbolic analysis of the construction of Hamiltonians for AQC. In Chapter 5, we give a general Hamiltonian construction for the pseudo-Boolean optimization problem. In Chapter 6, we show an alternative construction of local Hamiltonians based on a study of graph decompositions and a dynamic programming approach. Finally, in Chapter 7, we present conclusions.

William Cruz-Santos and Guillermo Morales-Luna
October 2014

Acknowledgments

This book is the result of an exciting journey through computer science and physics, and it would not have been possible without the contributions of many people. We would like to thank Dr. Debrup Chakraborty and Dr. Sergio Víctor Chapa from the Computer Science Department at Cinvestav-IPN, Dr. Micho Durdevich Lucich from Mathematics Institute at UNAM, and Dr. Alán Aspuru-Guzik from Harvard University, whose valuable comments helped to improve this work. We would like to mention professors Valentina Harizanov and Joe Mourad from GWU for supporting a research stay at the Center for Quantum Computing, Information, Logic and Topology in Washington DC and for their valuable advice.

We would also like to thank Dr. Salvador E. Venegas Andraca for his support and sharing his interesting point of view of science. We thank Professor Marco Lanzagorta for encouraging us to present this book on adiabatic quantum computing. We also want to thank Professor Jeffrey K. Uhlmann who, together with Professor Lanzagorta, are the editors of Morgan & Claypool's *Synthesis Lectures on Quantum Computing*. Furthermore, we want to thank Mr. Michael B. Morgan, President of Morgan & Claypool Publishers, for supporting this project.

Finally, but no less important, we want to thank the Cinvestav-IPN for its support and pleasant stay at their facility. We want to finally acknowledge the support from Mexican Conacyt.

William Cruz-Santos and Guillermo Morales-Luna
October 2014

CHAPTER 1

Introduction

The first idea for performing computations using a quantum computer was proposed by Richard Feynman in 1982. He wondered whether it was possible to simulate a quantum system by means of a universal quantum simulator [39]. The Feynman's proposal was the initial motivation for many future experimental and theoretical results in the field of *Quantum Computation* (QC).

The first important theoretical result was given in [87]. It proposed a polynomial time quantum algorithm to the problem of factoring an integer number into its prime factors, something thought to be very difficult. Since then, many quantum algorithms were proposed; for instance, in [46] a sublinear time quantum algorithm is given to solve the problem of finding an element in a non-structured database.

Quantum algorithms (QA) are based on the application of unitary operators that act in a finite dimensional Hilbert space. Thus, a QA consists of consecutive applications of unitary operators over an initial quantum state, and the output of the algorithm is obtained by performing a *quantum measurement* over the final state. This approach is known as the *quantum circuit model* (QCM) [70].

On the other hand, *Adiabatic Quantum Computation* (AQC) was introduced in [38] and it has been applied to solve optimization problems. It is based on the construction of a time-dependent Hamiltonian which codify the optimal solution of the given optimization problem into its ground state. AQC makes use of the *Adiabatic Theorem* [45, 65] to approximate solutions of the Schrödinger equation in which a slow evolution occurs.

Although AQC approach is defined by the solutions of the continuous Schrödinger equation, it was proved that AQC is equivalent to QCM [2], and therefore AQC is a universal model of computation. A very active area in AQC deals with the problem to determine a time lower-bound that an AQC algorithm requires in order to obtain an optimal solution. Theoretical and experimental results to this problem were presented in [38] to the 3-SAT problem.

The Hamiltonian operators used in AQC should be local for convenience. *Local Hamiltonian* operators are expressed as a polynomial sum, i.e., the addition of a polynomial number of Hamiltonians each acting over a reduced number of states. With the use of local Hamiltonians, it is possible to perform computations in a local way, affecting only a neighborhood of states in the quantum system. A related important problem is the *Local Hamiltonian Problem* (LHP) that consists in deciding if a given Hamiltonian acting in a Hilbert space of dimension n has an eigenvalue below a, or if all its eigenvalues are at less b, where a and b are real numbers such that $b - a \geq n^{-O(1)}$.

The LHP is known to be QMA-complete [21, 56–58] where QMA is the class of problems that can be solved in polynomial time by QA's. From the point of view of complexity theory, the LHP can be seen as the quantum analog of the SAT problem for the class of problems NP and restricted versions of the LHP coincide with NP-complete problems [95].

In [1] a general technique was proposed to decompose a Hamiltonian into a polynomial sum of local Hamiltonians, based on the assumption that the given Hamiltonian is d-sparse and row-computable. The Hamiltonian decomposition is important to the *Hamiltonian simulation* which consists of computing the matrix exponentiation of a given Hermitian matrix that results into a unitary matrix [13, 23, 74].

Formally, the Adiabatic Theorem guaranties that the final ground state of a given Hamiltonian can be approximated with arbitrary accuracy, and assuming that QC cannot solve NP-complete problems, it follows that AQC is a means of obtaining an approximation to the ground state and the ground state energy. The ratio of this approximation and its dependence on the hardness of the problem are not well understood yet. On the other hand, classical algorithms have been proposed; for instance, in [10] it was shown a classical approximation algorithm for evaluating the ground-state energy of the classical *Ising Hamiltonian* with linear terms on an arbitrary planar graph. Also, a classical approximation algorithm is proposed to the LHP [71].

The current construction of local Hamiltonians does not use the structure of the given problem. For instance, in [36] an adiabatic quantum algorithm is given for the MAX-SAT problem. It is based on a natural equivalence between clauses and Hamiltonians defined for every literal in the given instance. A similar construction is given in [75] for the *protein folding problem*, it is based on the *Ising model* to describe the local interactions in a lattice.

The design and construction of Hamiltonians have important consequences in the running time and convergence for AQC algorithms [4, 25, 37, 91]. In this book we deal with the problem of local Hamiltonian construction for combinatorial optimization graph problems [14, 17, 18] and with the formulation of such problems into the Pseudo-Boolean optimization approach.

The main contributions in this book are as follows.

- A complete treatment and analysis of the AQC applied to the 3-SAT problem. We analyze the syntactical construction of the initial and final Hamiltonians involved in the AQC algorithm, such an analysis is useful in order to perform numerical simulations of the adiabatic algorithm, i.e., to avoid a direct construction of the Hamiltonians by means of tensor products. Also, we provide a precise description of the computational complexity of the AQC algorithm.

- A general model in terms of *pseudo-Boolean functions* to solve optimization problems. The main idea is to model any combinatorial optimization as a *quadratic pseudo-Boolean function* [20]; then a Hamiltonian operator is constructed such that its ground state corresponds to the point that minimizes the quadratic pseudo-Boolean function. We also show that any problem expressed in monadic second-order logic has an associated pseudo-Boolean function with a bounded number of variables.

- A dynamic programming approach to solve NP-hard problems. This is based on the Tree Decompositions of graphs introduced in [79–82] and a dynamic programming technique in which a graph problem can be decomposed into smallest subproblems and then composed in order to construct a global solution of the problem [14, 17]. Based on this decomposition, we propose the Hamiltonian construction for AQC on each subproblem of the tree decomposition and composed in a global Hamiltonian whose ground state is the point at which the pseudo-Boolean function has its minimum.

CHAPTER 2

Approximability of NP-hard Problems

The purpose of this chapter is to give a succinct introduction to computational complexity and its principal problems, the scope which ranges from complexity classes to approximating optimization problems. Background knowledge of *Turing machines* (TM) is assumed (see [7, 9, 42]).

Here, we emphasize the relationship between decision and optimization problems. In order to do this, we recall the class P in terms of *Deterministic Turing Machines* (DTM) and the class NP, in terms of DTMs that test membership in languages.

The class NP has a probabilistic characterization in terms of *Probabilistic Turing Machines* (PTM), captured in the PCP theorem. The PCP theorem has important applications in approximating solutions to NP-hard problems using a standard methodology.

We also introduce the randomized complexity classes and their connections with optimization problems. Randomized computation is related to quantum computation by its probabilistic nature; that is, the quantum complexity class BQP contains the classical complexity class BPP. In what follows, we make a survey of these notions.

2.1   BASIC DEFINITIONS

Let L be a language. There exists a DTM M that recognizes L whenever L is decidable. Given a DTM M, let L_M be the language recognized by M. M is said to be *polynomial-time*, if for each input string x, with $|x| = n$, M performs at most $O(n^k)$ computing steps for a fixed non-negative exponent k, where $|x|$ denote the length of the string x.

Definition 2.1   The class P consists of all languages recognized by polynomial time DTMs.

Let L be a language. A *verification procedure* for L is a DTM V that satisfies the following conditions.

1. *Completeness*: For each $x \in L$ there exists a string y such that $V(x, y) = 1$.
   (V accepts y as a valid proof for the membership of x in L)

2. *Soundness*: For each $x \notin L$ and every string y it holds that $V(x, y) = 0$.
   (V rejects y as proof for the membership of x in L)

A language L has an *efficiently verifiable proof system* if there exists a polynomial p and a polynomial time verification procedure V such that, for each $x \in L : (\exists y : |y| \leq p(|x|) \wedge V(x, y) = 1)$ and for every $x \notin L$ and every y, the equation $V(x, y) = 0$ holds.

Definition 2.2   The class NP consists of all languages that have efficient verifiable proof systems.

From Definitions 2.1 and 2.2 it is satisfied that $P \subseteq NP$.

Although the definition of the classes P and NP have been expressed in terms of decision problems, there exist equivalent definitions for search problems.

A *search problem* is a relation $R \subseteq \{0, 1\}^* \times \{0, 1\}^*$. For an instance x of R, let $R(x) := \{y : (x, y) \in R\}$ be the set of solutions of x. A function $f : \{0, 1\}^* \to \{0, 1\}^* \cup \{\bot\}$ solves the search problem R if for every x; whenever $R(x) \neq \emptyset$ we have $f(x) \in R(x)$, otherwise $f(x) = \bot$.

Let L_1, L_2 be two languages; L_1 is *reducible* to L_2 if there exists a function $\Phi : L_1 \to L_2$ such that $x \in L_1$ if and only if $\Phi(x) \in L_2$. The language L_1 is said to be *polynomially reducible* to L_2 if the reduction map Φ can be computed in polynomial time. In this case, it is written $L_1 \leq_p L_2$.

A language $L \in NP$ is NP-*complete* if for each $L' \in NP$, $L' \leq_p L$. The NP-complete problems are thus the most difficult problems in the class NP.

2.2   PROBABILISTIC PROOF SYSTEMS

A verification procedure given by a DTM M can be extended by changing M with a PTM that uses a source of random bits for its computation. From now on, a verification procedure will be called a *verifier* for short.

Definition 2.3   A verifier V is a PTM having an input tape, work tape, source of random bits, and read-only tape called proof string π. V has random access to π and the operation of reading a bit in π is called a query.

The source of random bits of a verifier can be viewed as an input random string ρ. Figure 2.1 shows a verifier and its components. It can be seen that a verifier is equivalent to a verification procedure when the verifier does not use the random string for its computation.

Let L be a language and $q, r : \mathbb{N} \to \mathbb{N}$ be two functions. L has a $(r(n), q(n))$-*restricted verifier* if there is a verifier V such that satisfies the following conditions.

1. *Efficiency*: For each input x with $|x| = n$ and given a proof string π of length at most $q(n)2^{r(n)}$, V uses at most $r(n)$ random bits and queries at most $q(n)$ positions in π. Then V outputs 1 for "accept" or outputs 0 for "reject."

2. *Completeness*: For each $x \in L$ there exists a proof π_x such that for every random string ρ, $\Pr[V(x, \pi_x, \rho) = 1] = 1$.

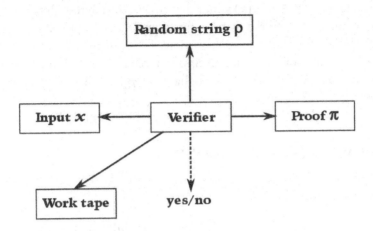

Figure 2.1: Probabilistic Turing machine (verifier).

3. *Soundness*: For each $x \notin L$ and every proof π and random string ρ, $\Pr[V(x, \pi, \rho) = 1] \leq \frac{1}{2}$.

Definition 2.4   The class $PCP[r(n), q(n)]$ consists of all languages that have $(r(n), q(n))$-restricted verifiers.

Note that $NP = PCP[0, \text{poly}(n)]$ where $\text{poly}(n) = \bigcup_{k \in \mathbb{N}} n^k$.

Theorem 2.5 Arora and Safra [8] $NP = PCP[\log n, 1]$.

Theorem 2.5 asserts that in order to check membership, just a constant number of accesses to the proof and a logarithmic number of random bits are required. Also, Theorem 2.5 has important applications to prove the hardness in approximating NP-hard problems [48, 90, 94], as we review in the next section.

2.3   OPTIMIZATION PROBLEMS

Definition 2.6   An optimization problem Π is a tuple $(I_\Pi, \text{sol}_\Pi, m_\Pi, \text{goal}_\Pi)$ where I_Π is the set of instances of Π, $\text{sol}_\Pi : I_\Pi \to \Omega(x)$ is a function that associates to any instance $x \in I_\Pi$ the set of feasible solutions of x, and $m_\Pi : I_\Pi \times \text{sol}_\Pi \to \mathbb{Z}^+$ is the measure function and $\text{goal}_\Pi \in \{\min, \max\}$.

The optimal solution to an instance $x \in I_\Pi$ is denoted as $y^*(x) \in \text{sol}_\Pi(x)$ according to goal_Π and its measure as $m_\Pi^*(x)$.

Let $\Pi = (I_\Pi, \text{sol}_\Pi, m_\Pi, \text{goal}_\Pi)$ be an optimization problem, Π is in the class NPO if the set of instances I_Π can be recognized in polynomial time, namely for each $x \in I_\Pi$ there exists a polynomial p and for any $y \in \text{sol}_\Pi(x)$ with $|y| \leq p(|x|)$, the membership of y in $\text{sol}_\Pi(x)$ can be decided in polynomial time and the measure function can be computed in polynomial time.

The class NPO is the optimization version of NP, in the sense that every optimization problem in NPO has its corresponding decision problem in NP. A problem Π is NP-hard if there exists an NP-complete problem Π' such that $\Pi' \leq_{T,p} \Pi$ where $\leq_{T,p}$ is a polynomial Turing reduction (See [9]).

The NP-hard problems are the most difficult problems in NPO.

2.3.1   APPROXIMATION ALGORITHMS

Let Π be an optimization problem; for any $x \in I_\Pi$ and for any value $y \in \text{sol}_\Pi(x)$ the *performance ratio* of y with respect to x is defined as:

$$R(x, y) = \max \left\{ \frac{m_\Pi(x, y)}{m_\Pi^*(x)}, \frac{m_\Pi^*(x)}{m_\Pi(x, y)} \right\}.$$

The performance ratio is always a number greater than or equal to 1 and is closer to 1 as y is closer to the optimum solution.

An algorithm T is an ε-*approximation algorithm* for Π if, given any $x \in I_\Pi$, $R(x, T(x)) \leq \varepsilon$, and it is said that Π is ε-approximated.

APX is the class of all NPO problems Π such that, for some $\varepsilon > 1$, there exists a polynomial time ε-approximation algorithm for Π.

Let Π be an NPO problem. An algorithm T is said to be an *approximation scheme* for Π if, for any $x \in I_\Pi$ and for any rational $\varepsilon > 1$, $T(x, \varepsilon)$ returns a feasible solution of x whose performance ratio is at most ε.

Definition 2.7   An NPO problem Π belongs to the class PTAS if it admits a polynomial-time approximation scheme.

Note that the time complexity of an approximation scheme may be of the type $2^{1/(\varepsilon-1)} p(|x|)$ or $|x|^{1/(\varepsilon-1)}$ where p is a polynomial.

An NPO problem Π belongs to the class FPTAS if it admits a fully polynomial-time approximation scheme, that is, an approximation scheme whose time complexity is bounded by $q(|x|, 1/(\varepsilon - 1))$ where q is a polynomial.

Clearly, FPTAS $\subseteq$ PTAS $\subseteq$ APX $\subseteq$ NPO.

2.4   RANDOMIZED COMPLEXITY CLASSES

In Section 2.2 a PTM was introduced and considered as a verifier for membership in languages. Here a more complete discussion is given in order to introduce the complexity classes of problems that can be solved in terms of polynomial randomized algorithms.

First, a formal definition of a PTM is given.

Definition 2.8   Probabilistic polynomial time machine   It is said that M is a probabilistic polynomial time machine if there exists a polynomial p such that when invoked on any input $x \in \{0, 1\}^*$, machine M always halts within at most $p(|x|)$ steps (regardless of the outcome of its internal coin tosses). In such a case, $M(x)$ is a random variable.

An alternative definition was given in Section 2.2, in terms of the languages that can be recognized by PTMs. This definition is the following.

Let L be a language. It is said that L has a PTM M if there exists a polynomial p such that for each $x \in L$, $M(x)$ can be computed within at most $p(|x|)$ steps. In such a case, $M(x)$ is a random variable over the output distribution of M with input x.

One important aspect of probabilistic polynomial time machines is that on any input $x \in \{0, 1\}^*$, a PTM M may fail to give an answer or equivalently, the machine M may fail to test membership for a language. That is, with some specified probability, these machines may fail to produce the desired output.

One aspect of the type of failure of a PTM is whether, in case of failure, the algorithm produces a wrong answer or merely an indication that it failed to find a correct answer. Let us consider three types of failure, giving rise to three different types of algorithms.

1. *Two-sided error.* This is the most liberal notion of failure, and it is originated from the setting of decision problems, where it means that the algorithm may err in both directions (i.e., it may rule that a yes-instance is a no-instance, and vice versa).

2. *One-sided error.* This is an intermediate notion of failure, and again it is originated from the setting of decision problems, where it means that the algorithm may err only in one direction (i.e., either on yes-instance or no-instance).

3. *Zero-sided error.* This is the most conservative notion of failure. In this case, the algorithm's failure amounts to indicating its failure to find an answer (by outputting a special symbol). That means that the algorithm never provides a wrong answer.

Before continuing it is important to mention another aspect of the probabilistic polynomial algorithms. According to the considered types of failure, an algorithm may fail to give a correct answer with certain probability. This error probability in practice must be *negligible*, which intuitively means that the failure event is so rare that it can be ignored in practice. Formally, a function is negligible, with respect of the relevant parameter, vanishes faster than the reciprocal of any positive polynomial. In other words, a function $\mu : \mathbb{N} \to \mathbb{R}$ is negligible, if for every positive polynomial $poly(\cdot)$ there exits an integer N_{poly} such that for all $x > N_{poly}$,

$$|\mu(x)| < \frac{1}{poly(x)}.$$

In the following, the error probability of a PTM is considered a negligible function with respect to the relevant parameter.

2.4.1 THE COMPLEXITY CLASS BPP

In this section the complexity class BPP is introduced that correspond to languages that can be recognized by algorithms that may fail in both directions (two-sided error) with negligible probability. The latter requirement guarantees the usefulness of such algorithms, because the error probability can be ignored.

It is said that a probabilistic (polynomial-time) algorithm A decides membership in S if for every x it is satisfied that $\Pr[A(x) = \chi_S(x)] > 1 - \mu(|x|)$, where χ_S is the characteristic function of S (i.e., $\chi_S(x) = 1$ if $x \in S$ and $\chi_S(x) = 0$ otherwise) and μ is a negligible function. The class of decision problems that are solvable by probabilistic polynomial-time algorithms is denoted as BPP (Bounded-error Probabilistic Polynomial time).

A formal definition of the class BPP is the following.

Definition 2.9 The Complexity Class BPP   A decision problem S is in BPP if there exists a probabilistic polynomial-time algorithm A such that for every $x \in S$ it is satisfied that $\Pr[A(x) = 1] \geq \frac{2}{3}$ and for every $x \notin S$ it is satisfied that $\Pr[A(x) = 0] \geq \frac{2}{3}$.

The choice of the constant $\frac{2}{3}$ is immaterial, and any other constant greater than $\frac{1}{2}$ yields the very same class. The latter is due to the error reduction theorem [42], it states that invoking a probabilistic polynomial-time algorithm for an adequate number of times, and ruling by majority, its error probability can be reduced.

A clear result regarding the relation of the classes BPP and P is that $P \subseteq BPP$. On the other hand, it is an open problem whether $BPP = P$, it is commonly conjectured that the answer is negative. Another known result about the class BPP was given by the Sipser-Lautemann theorem which states that BPP is contained in the polynomial time hierarchy, a more specifically BPP $\subseteq \Sigma_2 \cap \Pi_2$.

2.4.2 THE COMPLEXITY CLASS RP

Here, the notions of probabilistic polynomial-time algorithms having one-sided error are considered. The notion of one-sided error refers to a natural partition of the set of instances into yes-instances vs. no-instances in the case of decision problems.

Let us give a formal definition of the class RP.

Definition 2.10 The Complexity Class RP   A decision problem S is in RP if there exists a probabilistic polynomial-time algorithm A such that for every $x \in S$ it is satisfied that $\Pr[A(x) = 1] \geq \frac{1}{2}$ and for every $x \notin S$ it is satisfied that $\Pr[A(x) = 0] \geq 1$.

Here, again the choice of the constant $\frac{1}{2}$ is immaterial, and any other constant greater than zero yields the very same class.

It can be proved that for every $S \in$ NP there exists a probabilistic polynomial-time algorithm A such that for every $x \in S$ it is satisfied that $\Pr[A(x) = 1] > 0$ and for every $x \notin S$ it is satisfied that $\Pr[A(x) = 0] = 1$. That is, A has error probability at most $1 - \exp(-poly(|x|))$ on yes-instances but never errs on no-instances. Thus, it means that RP ⊆ NP. Another result that can be proved by the error reduction theorem is the following containment RP ⊆ BPP.

The class coRP corresponds to the opposite direction of one-sided error probability, and is defined as coRP = {{0, 1}* \ S : S ∈ RP}. Let us give a formal definition.

Definition 2.11 The complexity class coRP A decision problem S is in coRP if there exists a probabilistic polynomial-time algorithm A such that for every $x \in S$ it is satisfied that $\Pr[A(x) = 1] = 1$ and for every $x \notin S$ it is satisfied that $\Pr[A(x) = 0] \geq \frac{1}{2}$.

It can be proved that BPP is reducible to coRP by one-sided error randomized Karp-reductions [42]. Moreover, BPP is trivially reducible to coRP by two-sided error randomized Karp-reductions.

2.4.3 THE COMPLEXITY CLASS ZPP

Finally, we consider probabilistic polynomial-time algorithms that never err, but may fail to provide an answer. The actual definition of the class ZPP (Zero-error Probabilistic Polynomial time) is in terms algorithms that output ⊥ (indicating failure) with probability at most $\frac{1}{2}$.

A formal definition of ZPP is the following.

Definition 2.12 The complexity class ZPP A decision problem S is in ZPP if there exists a probabilistic polynomial-time algorithm A such that for every $x \in \{0, 1\}^*$ it is satisfied that $\Pr[A(x) \in \{\chi_S(x), \perp\}] = 1$ and $\Pr[A(x) = \chi_S(x)] \geq \frac{1}{2}$, where $\chi_S(x) = 1$ if $x \in S$ and $\chi_S(x) = 0$ otherwise.

Here again the choice of the constant $\frac{1}{2}$ is immaterial and any constant greater than zero yields the very same class. An alternative definition of the class ZPP is the following.

It is said that a decision problem S is solvable in *expected probabilistic polynomial time* if there exists a randomized algorithm A and a polynomial p such that for every $x \in \{0, 1\}^*$ it is satisfied that $\Pr[A(x) = \chi_S(x)] = 1$ and the expected number of steps taken by $A(x)$ is at most $p(|x|)$. Thus, we have the following.

Definition 2.13 Alternative definition of ZPP A decision problem S is in ZPP if and only if S is solvable in expected probabilistic polynomial time.

The class ZPP can be described in terms of the classes RP and coRP as follows ZPP = RP ∩ coRP [42].

Table 2.1 summarizes the complexity classes of the three types of failure for a language L that can be recognized by probabilistic polynomial-time algorithms.

From the point of view of optimization, a PTM can approximate solutions of NP-hard problems. For instance let $\Pi = (I_\Pi, \text{sol}_\Pi, m_\Pi, \text{goal}_\Pi)$ be an optimization problem, and for any instance $x \in I_\Pi$ and an integer $k \in \mathbb{Z}^+$ let us consider the decision problem Π_D with respect to

Table 2.1: Randomized class of languages

Type of error	Class	$x \in L$	$x \notin L$	$\perp$
Two-sided	BPP	$\Pr[M(x) = 1] \geq \frac{2}{3}$	$\Pr[M(x) = 0] \geq \frac{2}{3}$	
One-sided	RP	$\Pr[M(x) = 1] \geq \frac{1}{2}$	$\Pr[M(x) = 0] = 1$	
	coRP	$\Pr[M(x) = 1] = 1$	$\Pr[M(x) = 0] \geq \frac{1}{2}$	
Zero-sided	ZPP	$\Pr[M(x) = 1] = 1$	$\Pr[M(x) = 0] = 1$	$\Pr[M(x) = \perp] = 1$
		$\Pr[M(x) = 1] \geq \frac{1}{2}$	$\Pr[M(x) = 0] \geq \frac{1}{2}$	

Π, such that it is required decide whether $m^*_\Pi(x) \geq k$ if $\text{goal}_\Pi = MAX$ or whether $m^*_\Pi(x) \leq k$ if $\text{goal}_\Pi = \min$.

Thus, the language with respect to Π_D is $L_\Pi = \{(x,k) | x \in I_\Pi \wedge m^*_\Pi(x) \geq k\}$ if $\text{goal}_\Pi = \max$, and a PTM can be used to decide membership in this language with certain error probability (that is considered negligible).

2.4.4   QUANTUM COMPLEXITY

Quantum computation is realized in finite dimensional Hilbert spaces, the operations are realized as unitary operators over unit vectors represented as linear combinations of vectors in an orthonormal basis. Quantum measurements are the operations of reading the results. The notion of Quantum algorithms were first introduced in [12] using *Quantum Turing Machines* (QTM), which extend the classical TM.

Formally, a QTM is a triplet (Σ, Q, δ) where Σ is a finite alphabet, Q is a finite set of states with an distinguished initial state q_0, a final state q_f, and δ a quantum transition function

$$\delta : Q \times \Sigma \rightarrow \tilde{\mathbb{C}}^{Q \times \Sigma \times \{L, R\}},$$

where L, R is a left or right displacement over the tape machine and $\tilde{\mathbb{C}}$ is a set of computable complex numbers within some precision.

The QTM has a two-way infinite tape of cells indexed by $\mathbb{Z}$ and a single red/write tape head that moves along the tape. Given a pair $(q,s) \in Q \times \Sigma$, δ associates a complex number $\alpha \in \tilde{\mathbb{C}}$ to (q,s), such that the absolute value of α is the probability that the QTM will be in the new configuration (q', s') performing a left or right displacement over the tape machine.

A QTM halts if reaches the final configuration with state q_f, and it is said to be polynomial time if it performs a polynomial number of states transitions. It is possible to define an inner-product $\mathbb{S}$ space over $\tilde{\mathbb{C}}$ as the space of configurations of a QTM with the Euclidean norm; in this way, a linear combination of states in $\mathbb{S}$ is a superposition of states of a QTM.

A QTM M recognizes exactly the language L, if for each $x \in L$, M accepts x with probability 1 and for each $x \notin L$, M rejects x with probability 1.

The class EQP consists of all languages recognized exactly by polynomial time QTMs.

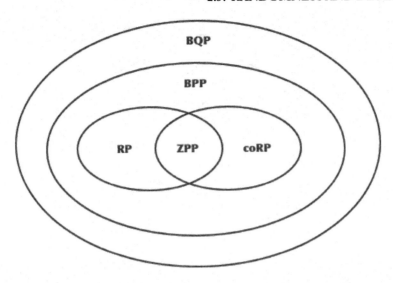

Figure 2.2: Contention of the class BQP with respect to the randomized classes BPP, RP, and ZPP.

A QTM M recognizes the language L with probability p if for each $x \in L$, M accepts x with probability p and for each $x \notin L$, M rejects x with probability $1 - p$.

Definition 2.14   The class BQP consists of all languages that are recognized by polynomial time QTM's with probability $\frac{2}{3}$.

The zero-sided error version of BQP is the class ZQP defined as follows: A language L is in ZQP if for every $x \in L$, x is accepted by some polynomial time QTM with probability $\frac{2}{3}$ and rejected with probability 0, and for every $x \notin L$, x is rejected by some polynomial time QTM with probability $\frac{2}{3}$ and accepted with probability 0.

Hence, $EQP \subseteq ZQP \subseteq BQP$.

The known relations with classical complexity classes are: $P \subseteq BQP$, $BPP \subseteq BQP$, and $BQP \subseteq PSPACE$ [70]. Figure 2.2 summarizes these class contentions.

Regarding the relation of the class QBP with respect to the class NP is an open problem, it is widely believed that BQP do not contain NP. Figure 2.3 shows this conjecture.

2.5   RANDOMNESS AND DETERMINISM

In this section we make a discussion about the need and usefulness of randomness in the design of algorithms, the *derandomization* of probabilistic polynomial-time algorithms is introduced (i.e., converting randomized algorithms to deterministic ones) and the derandomization of the entire class BPP is discussed.

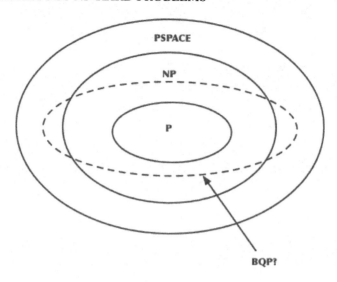

Figure 2.3: Possible contention of the class BQP with respect to the class NP.

In some cases, it is possible to obtain a deterministic polynomial-time algorithm from a probabilistic polynomial-time one with the same approximation rate. This has important consequences about the usefulness of randomness in the design of algorithms, if it is really buy us any advantage over the deterministic algorithms.

On the other hand, there are several notions regarding the nature of quantum algorithms, the most accepted one argues that the difficulty of predicting the distribution of probabilities of a quantum process is due to the lack of information about the experiment, or due to the limited accuracy of the measuring instruments. This is in accordance with the most recent theory of randomness that views randomness as an effect of an observer and thus as being relative to the observer's abilities (of analysis) [42]. Thus, randomness appears when any given distribution cannot be efficiently distinguished from the uniform distribution.

2.5.1 DERANDOMIZATION OF THE CLASS BPP

A probabilistic polynomial-time algorithm A has as input a random string provided on an additional input tape. This random string is used by the probabilistic algorithm for its internal computation, such that on any input $x \in \{0, 1\}^*$, $A(x)$ is a random variable. Of course, if for a fixed random string σ, and any input x, $A(\sigma, x)$ is not longer a random variable and a deterministic algorithm is obtained. A natural derandomization of a probabilistic polynomial-time algorithm is achieved if for any input $x \in \{0, 1\}^*$, all possible random strings of polynomial length are evaluated. In order to characterize this approach the input random-string of a probabilistic algorithm is replaced with a *pseudorandom string* (a sequence that cannot be told apart from the uniform distribution) generated with a *pseudorandom generator*.

Let us elaborate.

Definition 2.15   Pseudorandom generator   A deterministic polynomial-time algorithm G is called a pseudorandom generator if there exists a function $\ell : \mathbb{N} \to \mathbb{N}$ satisfying $\ell(k) > k$ for all k, such that for any probabilistic polynomial-time algorithm D, for any positive polynomial p, and for all sufficiently large k's, it is satisfied that

$$|\Pr[D(G(U_k)) = 1] - \Pr[D(U_{\ell(k)}) = 1]| < \frac{1}{p(k)},$$

where U_n denotes the uniform distribution over $\{0, 1\}^n$ and the probability is taken over U_k (resp., $U_{\ell(k)}$) as well as over the internal coin tosses of D.

A pseudorandom generator can be seen as a function that stretches k-bit short strings, called *seeds*, into $\ell(k)$-bit longer output sequences. Also, it must satisfied that any probabilistic algorithm D (*distinguisher*) cannot distinguish the output produced by the generator from a uniformly chosen sequence.

An application of a pseudorandom generator is the following: for any given probabilistic polynomial-time algorithm, replace the input random-string with a sequence generated by a pseudorandom generator. The following construction shows this substitution.

Construction 2.16   Let G be a pseudorandom generator with stretch function $\ell : \mathbb{N} \to \mathbb{N}$. Let A be a probabilistic polynomial-time algorithm, and $\rho : \mathbb{N} \to \mathbb{N}$ denote its randomness complexity. Denote by $A(x, r)$ the output of A on input x and coin tosses sequence $r \in \{0, 1\}^{\rho(|x|)}$. Consider the following randomized algorithm, denoted A_G.

On input x, set $k = k(|x|)$ to be the smallest integer such that $\ell(k) \geq \rho(|x|)$, uniformly select $s \in \{0, 1\}^k$, and output $A(x, r)$, where r is the $\rho(|x|)$-bit long prefix of $G(s)$.

That is, $A_G(x, s) = A(x, G'(s))$, for $|s| = k(|x|) = \arg\max_i\{\ell(i) \geq \rho(|x|)\}$, where $G'(s)$ is the $\rho(|x|)$-bit long prefix of $G(s)$.

Thus, using A_G instead of A, the randomness complexity is reduced from ρ to $\ell^{-1} \circ \rho$. For example, if $\ell(k) = k^2$, then the randomness complexity is reduced from ρ to $\sqrt{\rho}$.

In order to state the derandomization of the class BPP it is needed a more stronger notion of pseudorandom generator such that any polynomial-size circuits cannot distinguish the output produced by the generator from a uniformly chosen sequence.

Definition 2.17   Strong pseudorandom generator   A deterministic polynomial-time algorithm G is called a non-uniformly strong pseudorandom generator if there exists a stretch function, $\ell :$

$\mathbb{N} \rightarrow \mathbb{N}$, such that for any family $\{C_k\}_{k \in \mathbb{N}}$ of polynomial-size circuits, for any positive polynomial p, and for all sufficiently large k's

$$|\Pr[C_k(G(U_k)) = 1] - \Pr[C_k(U_{\ell(k)}) = 1]| < \frac{1}{p(k)}.$$

Using such strong pseudorandom generators, the class BPP can be derandomized as stated in the following theorem.

Theorem 2.18 [42] If there exists non-uniformly strong pseudorandom generators then BPP is contained in $\bigcap_{\epsilon > 0} \text{DTIME}(t_\epsilon)$, where $\text{DTIME}(t_\epsilon)$ is the class of decision problems that are solvable in time complexity $t_\epsilon(n) := 2^{n^\epsilon}$.

In the proof of Theorem 2.18 [42], a pseudorandom generator is used to shrink the randomness complexity of a BPP-algorithm, and then derandomization is achieved by scanning all possible seeds to this generator. Here, the pseudorandom generator is able to run in time that is exponential in its seed length. A more stronger result presented in [42] the pseudorandom generator is allowed to run in time that is polynomial in its seed length.

2.5.2 DERANDOMIZATION TECHNIQUES

There are several methods that allow in some cases to transform (derandomization) randomized algorithms into deterministic algorithms that work in a similar way as the given randomized algorithms. Two methods are considered; the method of *conditional probabilities* and the method of *reduction of the probability space*.

The method of conditional probabilities

Let Π be an optimization problem that w.l.g. assume that is a maximization problem, and let A be a probabilistic algorithm for Π. The random coin-tosses of algorithm A can be described by a random string $\sigma = (\sigma_i)_{i=0}^{n-1}$ on input x, where n is polynomially bounded with respect to the length of x. Each σ_i takes the value 1 with probability p_i and the value 0 with probability $1 - p$. Let Z be the output of algorithm A on any input x, thus Z is a random variable. Our goal is to construct a deterministic polynomial-time algorithm which computes an output with value at least $E[Z]$.

Remember that if the random bits $\sigma_0, \ldots, \sigma_{n-1}$ are fixed to constants $d_0, \ldots, d_{n-1}$, then algorithm A becomes a deterministic one. It can be proved that there is always a choice for $d_0, \ldots, d_{n-1}$ so that the output of this deterministic algorithm is at least $E[Z]$. The problem is to choose $d_0, \ldots, d_{n-1}$ deterministically. For instance, the choice of d_i depends on $d_0, \ldots, d_{i-1}$ that are chosen before. For a choice $d_0, \ldots, d_{i-1}$ for $\sigma_0, \ldots, \sigma_{i-1}$, a weight $w(d_0, \ldots, d_{i-1})$ is associated, defined as the conditional expectation $E[Z | \sigma_0 = d_0, \ldots, \sigma_{i-1} = d_{i-1}]$. The derandomization by the method of conditional probabilities is based on the assumption that the weights $w(d_0, \ldots, d_{i-1})$ can be computed in polynomial time.

The derandomization algorithm can be summarized as follows.

1. For $i = 0$ to $n - 1$ do

  (a) If $w(d_0, \ldots, d_{i-2}, 0) > w(d_0, \ldots, d_{i-2}, 1)$ then $d_{i-1} := 0$; else $d_{i-1} := 1$;

2. Output the result of A with the choices $d_0, \ldots, d_{n-1}$.

Theorem 2.19 [64] If all weights $w(d_0, \ldots, d_{i-1})$ can be computed in polynomial-time, then the derandomization algorithm computes deterministically in polynomial time a solution of Π with a value at least $E[Z]$.

In [52] approximation algorithms for MAXSAT and MAXLINEQ3-2 are shown, and was obtained applying the method of conditional probabilities. Similar results are obtained in [78] for the *Vector Selection problem*.

Reduction of the probability space

Let A be a probabilistic polynomial-time algorithm for some optimization problem. If there is only a polynomial number of possible random choices for this algorithm, then it is easy to obtain a deterministic polynomial-time algorithm from A. When analyzing a probabilistic algorithm one usually assumes that the random choices are mutually independent. But sometimes it is possible to analyze the probabilistic algorithm without the assumption of mutually independent random choices. For instance, it may suffice for the analysis to assume pairwise or d-wise independence.

Our goal is to obtain a probability space (Ω, P) and d-wise independent random variables where the sample space Ω has only polynomial size.

Let us consider the following definition.

Definition 2.20 d-wise independence The random variables $X_1, \ldots, X_n$ are called d-wise independent if each subset consisting of d of these random variables is independent, i.e., if for all $1 \leq i_1 < \cdots < i_d \leq n$ and for all $x_1, \ldots, x_d$ it is satisfied that

$$\Pr[X_{i_1} = x_1, \ldots, X_{i_d} = x_d] = \prod_{k=1}^{d} \Pr[X_{i_k} = x_k].$$

It is assumed that the random variables X_i only takes the two values 0 and 1.

Lemma 2.21 [3] *Let $X_1, \ldots, X_n$ be random variables. Let q be a prime power such that $q \geq n$ and $d \geq 1$. Then there is a uniform probability space (Ω, P), where $|\Omega| = q^d$, and d-wise independent* *random variables* $X'_1, \ldots, X'_n$ *over* (Ω, P) *and it is satisfied that for each* $i \in \{1, \ldots, n\}$,

$$|Pr[X'_i = 1] - Pr[X_i = 1]| \leq \frac{1}{2q}.$$

Lemma 2.21 can be used for the derandomization of some probabilistic algorithms. In [66] an approximation algorithm for the *Maximal Independent Set problem* using the reduction of the probability space method.

A discussion of the foregoing theory on derandomization is placed in the following paragraph.

The derandomization of the entire class BPP with respect to a strong pseudorandom generator was shown, where the random complexity of the corresponding deterministic algorithm is exponential in its seed length. But, it can be obtained a derandomization such that the generator is allowed to run in time that is polynomial in its seed length. On the other hand, it is possible to obtain a deterministic algorithm from a probabilistic algorithm with the same approximation rate, whenever it is possible reduce the probability space of random choices in a given probabilistic algorithm. Of course, the design of a probabilistic algorithm is much easier than the design of a deterministic algorithm, this advantage is exploited in computer science for the solution of difficult problems. But, conceptually, in some cases, these probabilistic polynomial-time algorithms have equivalent deterministic algorithms with the same behavior. Now, regarding quantum algorithms, it is important to consider its classical simulation by means of probabilistic polynomial-time algorithms. That is, for a given quantum algorithm belonging in BQP, we want to construct a probabilistic polynomial-time algorithm with a probability distribution close to that of the quantum algorithm. For instance, in [54] it was shown efficient simulations of quantum circuits for the class Clifford circuits, and in [89] it was shown efficient classical simulations of anneling processes applied to combinatorial optimization problems.

CHAPTER 3

Adiabatic Quantum Computing

This chapter is a self-contained introduction to *adiabatic quantum computing* (AQC) as a general approach to approximate solutions to optimization problems.

The first part is dedicated to recalling the basic notions of *Hilbert spaces* and their metrics. The linear operators and their properties are introduced in the evolution of Quantum Systems using the Schrödinger picture.

Also, brief terminology used with *Quantum Mechanics* are given, in order to define *states*, *observables*, *measurements*, and *dynamics* of a quantum system. An important part of this chapter is the exposition of the *Adiabatic Theorem*, which is the fundamental tool in AQC. We sketch the general algorithm for AQC to solve optimization problems [38].

Finally, we analyze the conditions in which the Adiabatic Theorem is satisfied and the influence of the geometric Berry phases in the evolution of the adiabatic paths [45].

3.1 BASIC DEFINITIONS

A *metric space* is a pair (M, d) where M is a non-empty set and $d : M \times M \to \mathbb{R}^+$ is a metric on M satisfying $\forall x, y, z \in M$:

1. $d(x, y) \geq 0$, and $d(x, y) = 0$ if and only if $x = y$,

2. $d(x, y) = d(y, x)$, and

3. $d(x, y) \leq d(x, z) + d(z, y)$.

A metric space M is *complete* if every Cauchy sequence converges in M.

If V is any vector space then, the elements of V will be written using bold lowercase letters as $\mathbf{x}, \mathbf{y}, \mathbf{z}$. Any complex vector space V with induced norm by the inner product is a metric space with metric defined as $d(\mathbf{x}, \mathbf{y}) = \|\mathbf{x} - \mathbf{y}\|$ for all $\mathbf{x}, \mathbf{y} \in V$.

Let $\langle \cdot | \cdot \rangle : V \times V \to \mathbb{C}$ be an inner product thus, for all $\mathbf{x}, \mathbf{y}, \mathbf{z} \in V, a, b \in \mathbb{C}$:

1. $\langle \mathbf{x} | \mathbf{y} \rangle \geq 0$ and the following holds $\langle \mathbf{x} | \mathbf{y} \rangle = 0$ if and only if $\mathbf{x} = \mathbf{y}$,

2. $\langle \mathbf{x} | a\mathbf{y} + b\mathbf{z} \rangle = a \langle \mathbf{x} | \mathbf{y} \rangle + b \langle \mathbf{x} | \mathbf{z} \rangle$, and

3. $\langle \mathbf{x} | \mathbf{y} \rangle = \langle \mathbf{y} | \mathbf{x} \rangle^*$.

The *induced norm* of a complex vector space V is defined as $\|\mathbf{x}\| = \langle \mathbf{x} | \mathbf{x} \rangle^{\frac{1}{2}}$ with $\mathbf{x} \in V$, and for any $\mathbf{x}, \mathbf{y} \in V, a \in \mathbb{C}$,

1. $\|\mathbf{x}\| \geq 0$, and $\|\mathbf{x}\| = 0$ if $\mathbf{x} = \mathbf{0}$,

2. $\|\mathbf{x} + \mathbf{y}\| \leq \|\mathbf{x}\| + \|\mathbf{y}\|$, and

3. $\|a\mathbf{x}\| = |a|\|\mathbf{x}\|$.

Definition 3.1   A Hilbert space is a complex vector space $\mathbb{H}$ with inner product which is complete with respect to the metric induced by the inner product.

Let $\mathbb{H}_1 = \mathbb{C}^2$ be the complex Hilbert space of dimension 2. Let, for each $n > 1$, $\mathbb{H}_n = \mathbb{H}_{n-1} \otimes \mathbb{H}_1$ be the n-fold tensor product of $\mathbb{H}_1$. $\mathbb{H}_n$ is a Hilbert space of dimension $N = 2^n$, and for any $\mathbf{x} \in \mathbb{H}_n : \mathbf{x} = (x_0, \ldots, x_{N-1})$ is a vector with N complex entries.

For any two integers $i, j \in \mathbb{N}$, $i \leq j$, let $[\![i, j]\!]$ denote the collection of integers ranging from i to j, $[\![i, j]\!] = \{i, i+1, \ldots, j-1, j\}$.

Let $\langle \cdot | \cdot \rangle : \mathbb{H}_n \times \mathbb{H}_n \to \mathbb{C}$ be the inner product in $\mathbb{H}_n$ defined as:

$$\forall \mathbf{x}, \mathbf{y} \in \mathbb{H}_n : \langle \mathbf{y} | \mathbf{x} \rangle = \sum_{i=0}^{N-1} y_i^* x_i = (\mathbf{y})^H \mathbf{x},$$

where $(\mathbf{y})^H = (\mathbf{y}^T)^*$ is the *Adjoint Hermitian* of $\mathbf{y}$.

A vector $\mathbf{x} \in \mathbb{H}_n$ is a *unit vector* if $\|\mathbf{x}\| = 1$. A basis for $\mathbb{H}_n$ is a linearly independent vector family $(\mathbf{x}_i)_{i=0}^{N-1}$ satisfying $\forall \mathbf{z} \in \mathbb{H}_n : \mathbf{z} = \sum_{i=0}^{N-1} \alpha_i \mathbf{x}_i$ for some complex numbers $(\alpha_i)_{i=0}^{N-1}$. A basis $(\mathbf{x}_i)_{i=0}^{N-1}$ is orthonormal if, for all $i, j \in [\![0, N-1]\!]$ with $i \neq j$, $\langle \mathbf{x}_i | \mathbf{x}_j \rangle = 0$.

3.1.1   LINEAR OPERATORS

Let $\mathbb{H}_n$ be a Hilbert space and let $T : \mathbb{H}_n \to \mathbb{H}_n$ be an operator, T is a linear operator if $T(\sum_{i=0}^{k-1} a_i \mathbf{x}_i) = \sum_{i=0}^{k-1} a_i T(\mathbf{x}_i)$ where $\mathbf{x}_i \in \mathbb{H}_n$ and $a_i \in \mathbb{C}$ for all $i \in [\![0, k-1]\!]$. Let $I : \mathbb{H}_n \to \mathbb{H}_n$ be the *identity operator* and $0 : \mathbb{H}_n \to \mathbb{H}_n$ be the *zero operator*: for any $\mathbf{x} \in \mathbb{H}_n$, $I\mathbf{x} = \mathbf{x}$ and $0\mathbf{x} = \mathbf{0}$.

The set of all linear operators from $\mathbb{H}_n$ to $\mathbb{H}_n$ is denoted as $\mathcal{L}(\mathbb{H}_n)$. A linear operator $T \in \mathcal{L}(\mathbb{H}_n)$ is *self-adjoint* or *Hermitian* if $\langle T\mathbf{x}|\mathbf{y}\rangle = \langle \mathbf{x}|T\mathbf{y}\rangle$ and is *unitary* if $\langle T\mathbf{x}|T\mathbf{y}\rangle = \langle \mathbf{x}|\mathbf{y}\rangle$ for every choice of $\mathbf{x}, \mathbf{y} \in \mathbb{H}_n$.

Let $\mathrm{GL}(\mathbb{H}_n) = \{T \in \mathcal{L}(\mathbb{H}_n) | \det T \neq 0\}$ be the set of all invertible linear operators in $\mathcal{L}(\mathbb{H}_n)$ and let $\mathrm{SU}(\mathbb{H}_n) = \{T \in \mathrm{GL}(\mathbb{H}_n) | |\det T| = 1\}$ be the set of all unitary operators in $\mathrm{GL}(\mathbb{H}_n)$.

Given an orthonormal basis $(\mathbf{x}_i)_{i=0}^{N-1}$ for $\mathbb{H}_n$ and $T \in \mathcal{L}(\mathbb{H}_n)$, then the matrix representation of T is a matrix $T_{ij} \in \mathbb{C}^{N \times N}$ with entries t_{ij}:

$$\forall i \in [\![0, N-1]\!] : T\mathbf{x}_i = \sum_{j=0}^{N-1} t_{ij} \mathbf{x}_j.$$

We will use the matrix representation of an operator when it is clear from the context.

A linear operator $T \in \mathcal{L}(\mathbb{H}_n)$ is *Hermitian* if $T^H = T$ and is *unitary* if $T^H T = I$.

Definition 3.2 Let $T, S \in \mathcal{L}(\mathbb{H}_n)$ be two Hermitian matrices; T and S commute if and only if $[S, T] \equiv ST - TS = \mathbf{0}$.

For any T, S Hermitian operators and $a \in \mathbb{C}$, the following properties are satisfied:

1. $(aT)^H = a^* T^H$,

2. $(T + S)^H = T^H + S^H$, and

3. $(TS)^H = S^H T^H$.

TS is Hermitian if and only if T and S commute.

A Hermitian operator $T \in \mathcal{L}(\mathbb{H}_n)$ is *positive definite* if $\mathbf{x}^H T \mathbf{x} > 0$ for any vector $\mathbf{x} \in \mathbb{H}_n$.

Let $T \in \mathcal{L}(\mathbb{H}_n)$ be a linear operator, a nonzero vector $\mathbf{x} \in \mathbb{H}_n$ is *invariant* under T if and only if there exists a constant $\lambda \in \mathbb{C}$ such that $T\mathbf{x} = \lambda \mathbf{x}$. The number λ is said to be an *eigenvalue* of T and the vector $\mathbf{x}$ is said to be an *eigenvector* of T.

Let $T \in \mathcal{L}(\mathbb{H}_n)$; the *null subspace* of T is defined as $\mathcal{N}(T) := \{\mathbf{x} \in \mathbb{H}_n | T\mathbf{x} = \mathbf{0}\}$. The *spectrum* of T is defined as $\Lambda(T) := \{\lambda \in \mathbb{C} | \mathcal{N}(T - \lambda I) \neq \{\mathbf{0}\}\}$, i.e., $\Lambda(T) = \{\lambda_1, \ldots, \lambda_m\}$ is the set of all distinct eigenvalues of T. For any $j \in [\![1, m]\!]$, let $\gamma_j \equiv \dim \mathcal{N}(T - \lambda_j I)$ be the dimension of the subspace spanned by the eigenvectors corresponding to the eigenvalue λ_j. An eigenvalue λ_j is called *non-degenerate* if $\gamma_j = 1$ and it is called *degenerate* if $\gamma_j > 1$.

An important measure on linear operators is the *spectral norm*, which is defined as follows: For any linear operator $T : \mathbb{H}_n \to \mathbb{H}_n$,

$$\|T\| = \sup_{\mathbf{x} \neq \mathbf{0}} \frac{\|T\mathbf{x}\|}{\|\mathbf{x}\|} = \max_{\|\mathbf{x}\| = 1} \|T\mathbf{x}\|.$$

The spectral norm satisfies the following properties: For any $T, S \in \mathcal{L}(\mathbb{H}_n)$

1. $\|TS\| \leq \|T\| \|S\|$,

2. $\|T^H\| = \|T\|$,

3. $\|T \otimes S\| = \|T\| \|S\|$, and

4. $\|T\| = 1$ if T is unitary.

If T is a Hermitian operator then its spectral norm $\|T\| = \max\{|\lambda| \, | \, \lambda \in \Lambda(T)\}$ and $\|T\|^2$ is the largest eigenvalue of the operator $T^H T$.

3.2   QUANTUM STATES AND EVOLUTION

In the following, a brief introduction of the concepts and terminology of *Quantum Mechanics* (QM) used in this book are given (see [65] for a complete treatment in QM).

The *Quantum Theory* is a mathematical model of the physical world. In order to specify this model it is necessary to define the following concepts: *states, observables, measurements,* and *dynamics.*

1. *States*: A state is a complete description of a physical system. In QM a state is a unitary vector in a Hilbert space. Thus, the class of states of a quantum system coincides with the unit sphere on a Hilbert space. For instance, let $(\mathbf{x}_i)_{i=0}^{N-1}$ be an orthonormal basis for $\mathbb{H}_n$, for any $\mathbf{z} \in \mathbb{H}_n : \mathbf{z} = \sum_{i=0}^{N-1} \alpha_i \mathbf{x}_i$ where $\alpha_i \in \mathbb{C}$ with $i \in [\![0, N-1]\!]$, if $\mathbf{z}$ is unitary then $\sum_{i=0}^{N-1} |\alpha_i| = 1$. The squares of the absolute value of the scalars $(\alpha_i)_{i=0}^{N-1}$ correspond to a probability distribution and the value $|\alpha_i|^2$ is the probability of being in the state $\mathbf{x}_i$ for $i \in [\![0, N-1]\!]$.

2. *Observables and measurements*: An observable is a property of a physical system that in principle can be measured. In QM, an observable is a Hermitian operator. Let us see how an observable M can be represented as a sum of projector matrices, also-called the *spectral representation*. Let $M \in \mathcal{L}(\mathbb{H}_n)$ be an observable and let $(\mathbf{x}_i)_{i=0}^{N-1}$ be an orthonormal basis for $\mathbb{H}_n$, M can be represented as:

$$M = \sum_{i=0}^{N-1} \lambda_i P_i,$$

where $P_i = \mathbf{x}_i \mathbf{x}_i^H$ is the orthogonal projection onto the subspace spanned by the eigenvector $\mathbf{x}_i$ that corresponds to the eigenvalue $\lambda_i \in \Lambda(M)$. For all $i, j \in [\![0, N-1]\!] : P_i P_j = \delta_{ij} P_i$, $P_i^H = P_i$ and $\sum_{i=0}^{N-1} P_i^H P_i = \mathrm{Id}_n$.

An eigenstate of an observable is called an *energy state* and its corresponding eigenvalue is called the *energy*. The lowest energy of an observable is known as the *ground energy* and its corresponding energy state is known as the *ground state*. For any two observables $M_1, M_2 \in \mathcal{L}(\mathbb{H}_n)$, $M_1 + M_2$ is also an observable, but $M_1 M_2$ is an observable if and only if M_1 and M_2 commute.

The probability of finding a system in the energy λ_i of an observable M is given by:

$$\Pr(\lambda_i) = \| P_i \mathbf{x} \|^2 = \mathbf{x} P_i \mathbf{x}^H,$$

where $\mathbf{x}$ is the quantum state prior to the measurement and $\sum_{i=0}^{N-1} \Pr(\lambda_i) = 1$. If the outcome of a measurement is λ_i for an observable M, then the quantum state right after the measurement becomes:

$$\mathbf{y} = \frac{P_i \mathbf{x}}{(\mathbf{x} P_i \mathbf{x}^H)^{\frac{1}{2}}}.$$

3. *Dynamics*: The time evolution of a quantum state is described by a Hermitian operator also-called a *Hamiltonian* of the system. In the *Schrödinger picture* of dynamics, the time evolution of a quantum system is governed by the Schrödinger equation. Let $H : \mathbb{R} \to \mathrm{GL}(\mathbb{H}_n)$ be a time-dependent Hamiltonian and let $\mathbf{x} : \mathbb{R} \to \mathbb{H}_n$ be a differentiable transformation in the interval $I \subset \mathbb{R}$, then the Schrödinger equation is:

$$\forall t \in I : \frac{d}{dt}\mathbf{x}(t) = -iH(t)\mathbf{x}(t),$$

and can be rewritten as a first-order equation in the infinitesimal quantity dt as:

$$\mathbf{x}(t + dt) = U(dt)\mathbf{x}(t),$$

where $U(dt) := \mathrm{Id}_n - iH(t)dt$, hence $U^H U = \mathrm{Id}_n$. U is unitary if H is a time-independent Hamiltonian.

3.3   THE ADIABATIC THEOREM

The adiabatic approximation is a standard method of quantum mechanics used to derive approximate solutions of the Schrödinger equation in the case of a slowly varying Hamiltonian. The adiabatic approximation works as follows.

Put a quantum system in its ground state. If the Hamiltonian varies slowly enough, then the quantum system will stay in a state close to the instantaneous ground state of the Hamiltonian as the time goes on (see [65]).

3.3.1   ADIABATIC EVOLUTION

Let $\mathbb{H}_n$ be a Hilbert space and let $H : \mathbb{R} \to \mathrm{GL}(\mathbb{H}_n)$ be a time-dependent Hamiltonian. The differentiable transformation $\mathbf{x} : \mathbb{R} \to \mathbb{H}_n$ is a solution of the Schrödinger equation in the interval $I \subset \mathbb{R}$ if

$$\forall i \in I : i\frac{d}{dt}\mathbf{x}(t) = H(t)\mathbf{x}(t). \tag{3.1}$$

Let $J \subset \mathbb{R}$ be an interval and let $\tau : s \mapsto t = as + b$ be an affine transformation $J \to I$. Let $G : J \to \mathrm{GL}(\mathbb{H}_n)$ be such that $G(s) = aH(\tau(s))$.

Thus, if $\mathbf{x} : \mathbb{R} \to \mathbb{H}_n$ is a solution of (3.1) then

$$\forall s \in J : H(\tau(s))\mathbf{x}(\tau(s)) = i\frac{d}{dt}\mathbf{x}(\tau(s)) = i\frac{1}{a}\frac{d}{ds}\mathbf{x}(\tau(s)),$$

thus,

$$\forall s \in J : i\frac{d}{dt}\mathbf{x}(\tau(s)) = G(s)\mathbf{x}(\tau(s))$$

and $\mathbf{x} \circ \tau$ is a solution of the Schrödinger equation in J for the Hamiltonian $G = aH \circ \tau$. G is a continuous path in the space of Hermitian operators on $\mathbb{H}_n$.

For instance, if $J_{t_0} = [0, t_0]$ and $I = [0, 1]$ the affine transformation is $s \mapsto as + b = \frac{s}{t_0}$ and the Hamiltonian on J_{t_0} is $H_{t_0}(s) = \frac{1}{t_0} H(\frac{s}{t_0})$.

Let $\mathbf{x}_{t_0} : J_{t_0} \to \mathbb{H}_n$ be a solution of the equation

$$\forall s \in J_{t_0} : i \frac{d}{dt} \mathbf{x}_{t_0}(s) = H_{t_0}(s) \mathbf{x}_{t_0}(s). \tag{3.2}$$

Let $\{\lambda_0, \ldots, \lambda_{N-1}\} \subset \mathbb{R}^I$ be the spectrum of the Hamiltonian H such that for all $j \in [\![0, N-1]\!]$ and for all $t \in I$, there exists $\mathbf{y}_j(t) \in \mathbb{H}_n$ (instantaneous eigenstate of the Hamiltonian $H(t)$ with corresponding energy λ_j):

$$H(t) \mathbf{y}_j(t) = \lambda_j \mathbf{y}_j(t) \text{ with } \|\mathbf{y}_j(t)\| = 1$$

and

$$\lambda_0(t) \leq \cdots \leq \lambda_{N-1}(t).$$

The instantaneous eigenvalues are considered non-degenerated.

The path defined by the eigenvectors $(\mathbf{y}_0(t))_{t \in [0,1]}$ have extreme points $\mathbf{y}_0(0), \mathbf{y}_0(1)$. Let $\mathbf{z} \mapsto \langle \mathbf{y}_0(1) | \mathbf{z} \rangle$ be a linear transformation from $\mathbb{H}_n \to \mathbb{C}$ with respect to the instantaneous eigenvector $\mathbf{y}_0(1)$. If $\lambda_1(t) - \lambda_0(t) > 0$ for all $t \in [0, 1]$ then, the *Adiabatic Theorem* asserts that:

$$\lim_{t_0 \to +\infty} |\langle \mathbf{y}_0(1) | \mathbf{x}_{t_0}(t_0) \rangle| = 1.$$

This is the case of an infinitely slow or adiabatic passage. In other words, if the system is initially in an eigenstate of $H(0)$ it will, at time $t = 1$, under certain conditions to be specified later, have passed into the eigenstate of $H(1)$, that derives it by continuity.

An upper-bound for the time needed to satisfy the Adiabatic Theorem is the following:

$$T \geq \frac{\Delta_{max}}{\epsilon \delta_{min}^2},$$

where $\delta_{min} = \min_{0 \leq t \leq 1}(\lambda_1(t) - \lambda_0(t))$, $\Delta_{max} = \max \|\frac{d}{dt} H(t)\|$ and $\epsilon \in [0, 1]$ is the approximation ratio to the ground state of H.

3.3.2   QUANTUM COMPUTATION BY ADIABATIC EVOLUTION

The AQC was proposed in [38] as a general technique to solve optimization problems and was initially applied to the MAX-SAT problem and in [2] it was shown that AQC is equivalent to the circuit model of quantum computation and viceversa.

The adiabatic evolution of a quantum system can be used to solve optimization problems going from ground states to ground states of a time-dependent Hamiltonian.

Thus, given an optimization problem Π with its corresponding energy function or evaluation function and for a time-dependent Hamiltonian $H(t)$ for $0 \leq t \leq 1$. The ground state of

H at time $t = 1$ will correspond to the solution of the optimization problem. If the Hamiltonian H at time $t = 0$ is initially in an easily computable ground state (possibly in an uniform super-position of all basis states), then by the Adiabatic Theorem, for an infinitely slowly passage from $t = 0$ to $t = 1$, the evolution of the quantum system goes from the ground states to the ground states of H.

The general steps of the AQC algorithm are the following.

1. Prepare the quantum system in the ground state (which is known and easy to prepare) of another Hamiltonian H_0.

2. Encode the solution of an optimization problem into the ground state of a Hamiltonian H_f.

3. Evolve the quantum system slowly enough satisfying the Adiabatic Theorem with the Hamiltonian $H(t) = (1 - \frac{t}{T})H_0 + \frac{t}{T}H_f$ for a total time T. The final state $\mathbf{x}(t)$ at time $t = T$ will be (very close) the ground state of H_f (see Equation (3.1)).

4. Perform a measurement of the state $\mathbf{x}(t)$ at time $t = T$. With high probability the optimal solution of the optimization problem is found.

An important problem in AQC is to bound the time evolution T in order to satisfy the Adiabatic Theorem. Thus, for a given NP-hard problem, it still remains an open problem whether T could grows polynomially with respect to the size of the instance problem.

3.4   ADIABATIC PATHS

Let $\mathbb{H}_n$ be a Hilbert space of dimension $N = 2^n$ and let S be the unit sphere on $\mathbb{H}_n$, i.e., the class of all unitary states in $\mathbb{H}_n$. Let $t \mapsto H(t)$ be a continuous parametrization from $\mathbb{R}^+ \to \mathrm{GL}(\mathbb{H}_n)$ (a time dependent Hamiltonian). We claim that, for slowly changes of $H(t)$ in a closed internal, if the eigenvalue curves of H do not cross, then the instantaneous ground states remain invariants.

Let $t \mapsto \mathbf{x}(t)$ be a differentiable transformation and solution of the Schrödinger equation:

$$i\hbar \frac{d}{dt}\mathbf{x}(t) = H(t)\mathbf{x}(t). \tag{3.3}$$

Let $\Lambda(t) = \{\lambda_0(t), \ldots, \lambda_{N-1}(t)\}$ be the set of eigenvalues of $H(t)$ with $t \in \mathbb{R}^+$, sorted in decreasing order with respect to the absolute values. For each $j < N$, let $\mathbf{x}_j(t)$ be an eigenvector with corresponding eigenvalue $\lambda_j(t)$. Then:

$$H(t)\mathbf{x}_j(t) = \lambda_j(t)\mathbf{x}_j(t). \tag{3.4}$$

Assuming that the curves $\lambda_j(t)$ do not cross, i.e., each curve $\mathbf{x}_j : \mathbb{R}^+ \to S$ evolve adiabatically, the ground state of $H(t)$ is $\mathbf{x}_{N-1}(t)$.

At each time $t \in \mathbb{R}^+$ the set of eigenvectors $E(t) = (\mathbf{x}_j(t))_{j=0}^{N-1}$ is orthonormal in $\mathbb{H}_n$:

$$\forall k, j \in [\![0, N-1]\!] : \left[k \neq j \implies \mathbf{x}_k(t)^H \mathbf{x}_j(t) = \delta_{jk} \right].$$

Expressing a solution $\mathbf{x}(t)$ of Equation (3.3) as a linear combination of the elements in $E(t)$ modified by a phase factor:

$$\forall t \in \mathbb{R}^+ : \quad \mathbf{x}(t) = \sum_{j=0}^{N-1} c_j(t) \, e^{i\theta_j(t)} \, \mathbf{x}_j(t), \tag{3.5}$$

where each phase θ_j is given by

$$\forall t \in \mathbb{R}^+ : \quad \theta_j(t) = -\frac{1}{\hbar} \int_0^t \lambda_j(s) \, ds. \tag{3.6}$$

Then, according to Equation (3.3) and using elementary rules of derivation:

$$i\hbar \sum_{j=0}^{N-1} e^{i\theta_j(t)} \left[c_j'(t) \, \mathbf{x}_j(t) + c_j(t) \, \mathbf{x}_j'(t) + i\theta_j'(t) c_j(t) \, \mathbf{x}_j(t) \right] = \sum_{j=0}^{N-1} c_j(t) \, e^{i\theta_j(t)} \, H(t) \mathbf{x}_j(t). \tag{3.7}$$

and

$$c_k'(t) = -\sum_{j=0}^{N-1} c_j(t) e^{i(\theta_j(t) - \theta_k(t))} \, \mathbf{x}_k(t)^H \mathbf{x}_j'(t). \tag{3.8}$$

Now, deriving Equation (3.4), it follows:

$$H'(t) \, \mathbf{x}_j(t) + H(t) \, \mathbf{x}_j'(t) = \lambda_j'(t) \, \mathbf{x}_j(t) + \lambda_j(t) \, \mathbf{x}_j'(t),$$

hence

$$\mathbf{x}_k(t)^H H'(t) \, \mathbf{x}_j(t) +, \mathbf{x}_k(t)^H H(t) \, \mathbf{x}_j'(t) = \lambda_j'(t) \, \delta_{kj} + \lambda_j(t), \mathbf{x}_k(t)^H \mathbf{x}_j'(t).$$

Thus, since H is an adjoint operator,

$$k \neq j \implies \mathbf{x}_k(t)^H H'(t) \, \mathbf{x}_j(t) = (\lambda_j(t) - \lambda_k(t)), \mathbf{x}_k(t)^H \mathbf{x}_j'(t). \tag{3.9}$$

From (3.8) and (3.9), it follows:

$$c_k'(t) = -c_k(t) \, \mathbf{x}_k(t)^H \mathbf{x}_k'(t) - \sum_{j \in [\![0, N-1]\!] - \{k\}} c_j(t) \, \frac{e^{i(\theta_j(t) - \theta_k(t))}}{\lambda_j(t) - \lambda_k(t)} \, \mathbf{x}_k(t)^H H'(t) \, \mathbf{x}_j(t). \tag{3.10}$$

Now, if H change slowly enough with respect to t, then $\| H'(t) \|$ will be small, the terms in the right hand side of (3.10) are negligible, and the following approximation is found:

$$c_k'(t) = -c_k(t) \, \mathbf{x}_k(t)^H \mathbf{x}_k'(t),$$

whose solution is given by

$$\forall t \in \mathbb{R}^+ : \quad c_k(t) = c_k(0) e^{i \gamma_k(t)}. \tag{3.11}$$

where

$$\gamma_j(t) = i \int_0^t \mathbf{x}_k(s)^H \mathbf{x}_k'(s)\, ds. \tag{3.12}$$

Equation (3.5) can be written as

$$\forall t \in \mathbb{R}^+ : \quad \mathbf{x}(t) = \sum_{j=0}^{N-1} c_j(0) e^{i \gamma_j(t)} e^{i \theta_j(t)} \mathbf{x}_j(t). \tag{3.13}$$

If the system is initially in the ground state of $H(0)$, then $c_{N-1} = 1$ and $c_j = 0$ for any $j \neq N - 1$, therefore, from Equation (3.13),

$$\mathbf{x}(t) = e^{i \gamma_{N-1}(t)} e^{i \theta_{N-1}(t)} \mathbf{x}_{N-1}(t),$$

that is, the system remains in the same ground state up to a phase factor. The hypothesis concerning the not-crossing of the eigenvalue curves is used mainly in relation (3.10).

3.4.1 GEOMETRIC BERRY PHASES

In the following, we consider that the continuous parametrization $t \mapsto H(t)$ from $\mathbb{R}^+ \to \mathrm{GL}(\mathbb{H}_n)$ follows a closed trajectory.

Let $P \subset \mathbb{C}^k$ be a set of parameters, i.e., an open set in the topology of $\mathbb{C}^k$. Let $\mathbf{r} \mapsto H(\mathbf{r})$ be a continuous parametrization from $P \to \mathrm{GL}(\mathbb{H}_n)$. Let $t \mapsto \mathbf{r}(t)$ be a curve from $\mathbb{R}^+ \to P$ such that, for each t, k-parameters are selected. Let $t \mapsto \mathbf{x}(t)$ be a differentiable transformation and solution of the Schrödinger equation:

$$i\hbar \frac{d}{dt} \mathbf{x}(t) = H(\mathbf{r}(t)) \mathbf{x}(t). \tag{3.14}$$

Let $\Lambda(\mathbf{r}(t)) = \{\lambda_0(\mathbf{r}(t)), \ldots, \lambda_{N-1}(\mathbf{r}(t))\}$ be the set of eigenvalues of $H(\mathbf{r}(t))$ with $t \in \mathbb{R}^+$, sorted in decreasing order with respect to the absolute values. For each, $j < N$, let $\mathbf{x}_j(\mathbf{r}(t))$ be an eigenvector with corresponding eigenvalue $\lambda_j(\mathbf{r}(t))$. Then:

$$H(\mathbf{r}(t)) \mathbf{x}_j(\mathbf{r}(t)) = \lambda_j(\mathbf{r}(t)) \mathbf{x}_j(\mathbf{r}(t)). \tag{3.15}$$

Assuming that the curves $\lambda_j(\mathbf{r}(t))$ do not cross, i.e., the curve $\mathbf{r} : \mathbb{R}^+ \to P$ evolve adiabatically, the ground state of $H(\mathbf{r}(t))$ is $\mathbf{x}_{N-1}(\mathbf{r}(t))$.

Assuming that the solution $\mathbf{x}(t)$ of (3.14) coincide with the ground state up to a phase-shift factor:

$$\forall t \in \mathbb{R}^+ : \quad \mathbf{x}(t) = e^{i \phi_{N-1}(t)} \mathbf{x}_{N-1}(\mathbf{r}(t)). \tag{3.16}$$

By physical considerations, the dynamic phase factor is defined as:

$$\forall t \in \mathbb{R}^+ : \quad \theta_{N-1}(t) = -\frac{1}{\hbar} \int_0^t \lambda_{N-1}(\mathbf{r}(s)) \, ds. \tag{3.17}$$

The Berry phase is defined as the following difference:

$$\forall t \in \mathbb{R}^+ : \quad \gamma_{N-1}(t) = \phi_{N-1}(t) - \theta_{N-1}(t). \tag{3.18}$$

From Equations (3.14), (3.15), and (3.16), it follows:

$$\forall t \in \mathbb{R}^+ : \quad 0 = \frac{d}{dt} \mathbf{x}_{N-1}(\mathbf{r}(t)) + i \, \gamma'_{N-1}(t) \mathbf{x}_{N-1}(\mathbf{r}(t)). \tag{3.19}$$

Hence, $\forall t \in \mathbb{R}^+$:

$$
\begin{aligned}
\gamma'_{N-1}(t) &= -i \, i \gamma'_{N-1}(t) \mathbf{x}^H_{N-1}(\mathbf{r}(t)) \mathbf{x}_{N-1}(\mathbf{r}(t)) \\
&= i \, \mathbf{x}^H_{N-1}(\mathbf{r}(t)) \frac{d}{dt} \mathbf{x}_{N-1}(\mathbf{r}(t)) \\
&= i \, \mathbf{x}^H_{N-1}(\mathbf{r}(t)) D_{\mathbf{r}} \mathbf{x}_{N-1}(\mathbf{r}(t)) \frac{d}{dt} \mathbf{r}(t)
\end{aligned} \tag{3.20}
$$

(in the first equality the fact of unitarity was used, in the second the relation (3.19), and in the last the chain rule of derivation was used). Integrating Equation (3.20),

$$\gamma_{N-1}(t) = i \int_0^t \left(\mathbf{x}^H_{N-1}(\mathbf{r}(s)) D_{\mathbf{r}} \mathbf{x}_{N-1}(\mathbf{r}(s)) \right) \mathbf{r}'(s) \, ds \tag{3.21}$$

Assuming that the curve $\mathbf{r}$ is a circuit C in P i.e., for a time T, $\mathbf{r}(T) = \mathbf{r}(0)$ and $C = \{\mathbf{r}(t) | t \in [0, T]\}$, then Equation (3.21) becomes the so-called *Geometric Berry phase*:

$$\gamma_{N-1}(C) = i \oint_C \left(\mathbf{x}^H_{N-1}(\mathbf{r}) D_{\mathbf{r}} \mathbf{x}_{N-1}(\mathbf{r}) \right) d\mathbf{r}. \tag{3.22}$$

Since the states have constant length 1, it follows that:

$$
\begin{aligned}
0 &= D_{\mathbf{r}} \left(\mathbf{x}^H_{N-1}(\mathbf{r}) \mathbf{x}_{N-1}(\mathbf{r}) \right) \\
&= \left(D_{\mathbf{r}} \mathbf{x}^H_{N-1}(\mathbf{r}) \right) \mathbf{x}_{N-1}(\mathbf{r}) + \mathbf{x}^H_{N-1}(\mathbf{r}) D_{\mathbf{r}} \left(\mathbf{x}_{N-1}(\mathbf{r}) \right) \\
&= 2 \Re \left(\mathbf{x}^H_{N-1}(\mathbf{r}) D_{\mathbf{r}} \left(\mathbf{x}_{N-1}(\mathbf{r}) \right) \right),
\end{aligned}
$$

thus, the integrand $\mathbf{x}^H_{N-1}(\mathbf{r}) D_{\mathbf{r}} \mathbf{x}_{N-1}(\mathbf{r})$ in (3.22) is entirely imaginary and the geometric Berry phase $\gamma_{N-1}(C)$ is a real number. If $\gamma_{N-1}(C) = 0$, then the system is called *holonomic*. There are several types of non-holonomic systems and each one of them depends on the geometry where they belong.

Let us write (3.22) as:

$$\gamma_{N-1}(C) = \oint_C A_{N-1}(\mathbf{r}) \, d\mathbf{r}, \tag{3.23}$$

where

$$\mathbf{r} \mapsto A_{N-1}(\mathbf{r}) = i\mathbf{x}_{N-1}^{H}(\mathbf{r})D_{\mathbf{r}}\mathbf{x}_{N-1}(\mathbf{r}) \tag{3.24}$$

is a *recalibration potential* (*gauge potential*). A_{N-1} is *invariant under change of phases*; that is, given $\xi_{N-1} : P \to \mathbb{R}$ continuous, making a change of phase $\mathbf{y}_{N-1}(\mathbf{r}) = e^{i\,\xi_{N-1}(\mathbf{r})}\mathbf{x}_{N-1}(\mathbf{r})$, it results in $A_{N-1}(\mathbf{r}) = i\mathbf{y}_{N-1}^{H}(\mathbf{r})D_{\mathbf{r}}\mathbf{y}_{N-1}(\mathbf{r})$.

If the orthonormal basis $\left(\mathbf{x}_j(\mathbf{r})\right)_{j=0}^{N-1}$ for $H(\mathbf{r})$ is only changed by phases, then

$$\left(\mathbf{y}_j(\mathbf{r}) = e^{i\,\xi_j(\mathbf{r})}\mathbf{x}_j(\mathbf{r})\right)_{j=0}^{N-1},$$

the Berry phase remains invariant. Thus, the Berry phase is invariant under certain transformations $U(1)$ of Hamiltonians.

The *Aharanov-Bohm effect* appears when the Hamiltonian of a magnetic field and the corresponding electric field are moving in a circuit. The geometric Berry phase is not negligible, that is, the system is non-holonomic, and this produce the following effect: an electron beam that passes perpendicularly through a solenoid, forks, surrounding the solenoid to compose later. The relationship between the paths and adiabatic geometric phases is evident by the similarity of the phases involved. The relation (3.12) determines the phase involved in the evolution of an adiabatic path, while relation (3.21) determines properly the geometric Berry phase (3.22), when the path followed by a Hamiltonian is a circuit.

There are other geometric phases, such as the *Aharonov-Anandan*, *Pancharatnam*, or specific techniques such as NMR.

3.4.2   GEOMETRIC QUANTUM COMPUTATION

The Adiabatic Theorem provides an approximation to the instantaneous ground state of a Hamiltonian, but with the exception of a global phase, this phase can be divided into two parts: the dynamic phase and the geometric Berry phase; see Equations (3.17) and (3.18), respectively. The Berry phase depends only on the path taken, not on how fast the path is traversed. Hence, if we design a cyclic path of Hamiltonians, the Berry phase is totally determined.

Geometric Quantum Computing (GQC) [62, 67, 88, 98] uses the well-known fact in differential geometry that arises when a vector is transported parallel around a loop on a smooth manifold (see Figure 3.1). This vector may return rotated although there has no been local rotation along the loop. This global rotation is the *Holonomy* caused by the curvature of the underlaying space. In QM a state vector can be transported without locally rotating it around a loop in some quantum parameter space, and the resulting transformation has the same effect as applying a unitary matrix or phase factor that depends only on the global geometry of the loop.

In contrast to AQC that encodes the solution of an optimization problem into the ground state of a Hamiltonian, GQC encodes the solution of the problem into the Berry phase of the final state. In [99] shows an adiabatic algorithm for the *Counting Problem*, such that the solution is encoded in the Berry phase of the final state, rather than the ground state of the final state.

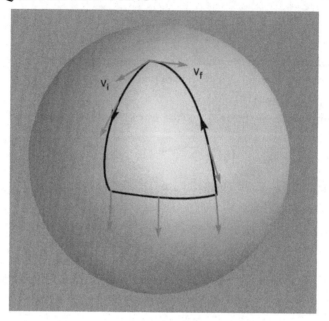

Figure 3.1: Parallel transport of a vector without local rotation on a curved surface (in a sphere). The final vector v_f has been rotated with respect to the initial vector v_i, and the rotation angle being the solid angle enclosed by the loop.

The final information is obtained by estimating the relative phase, rather than the usual quantum measurement.

There are several methods that have been proposed to build quantum gates based on geometric Berry phases. In [35] it was shown a geometric quantum computation scheme based on laser manipulation of a set of trapped ions. In [53] a controlled phase shift gate was proposed by performing a nuclear magnetic resonance experiment in which a conditional Berry phase is implemented. Since, the geometric Berry phase depends only on the geometry of the path executed, this suggest the possibility of an intrinsically fault-tolerant way of performing quantum gate operations (see also [72, 96]).

The GQC provides a scheme to design new quantum algorithms that are fault-tolerant for some kind of source of errors, but it depends on specific physical implementation such as NMR techniques. There is not yet a general technique to codify the solution of hard problems into the geometric Berry phase. An important problem is to propose a quantum algorithm based on the geometric Berry phase for the search problem in a database with time complexity equal to the proposed in [46].

In this book we do not consider the influence of the Berry phase in the adiabatic evolution, and in general we assume that the adiabatic paths follow an arbitrary trajectory.

CHAPTER 4

Efficient Hamiltonian Construction

In this chapter we propose a procedural construction of the Hamiltonian operators for AQC to avoid the direct tensor product construction. A complete treatment of AQC applied to the MAX-SAT problem is given, and the initial and final Hamiltonian are constructed in order to simulate AQC in a computationally efficient way [32].

The procedural construction of the Hamiltonian operators for AQC can be generalized for other optimization problems with a similar structure and it can be used in other applications such as Hamiltonian simulations and numerical analysis of the eigenvalue paths in the evolution of AQC.

4.1 AQC APPLIED TO THE MAX-SAT PROBLEM

In AQC, given a problem Π and any instance $\mathbf{x}$ of size n, a pair of Hamiltonians in a n-dimensional Hilbert space is determined. Each Hamiltonian corresponds to a Hermitian matrix represented by a $(2^n \times 2^n)$-complex matrix and in general corresponds to a sparse matrix. Two problems arise with the computational simulation of AQC: the first one is that the amount of memory required to store the two Hamiltonians becomes impractical for most of the actual computers, and the number of operations grows exponentially.

In order to deal with these two problems, we describe and characterize, in a procedural way, every entry of the initial and final Hamiltonians. Such a characterization of the Hamiltonians can reduce the amount of used memory to half.

Here we follow a SAT coding similar to the already standard codings [38, 49] into AQC. We will consider 3-SAT: The satisfiability decision problem for 3-clauses. And we provide a procedural construction of the initial and final Hamiltonian for the given instances.

4.1.1 SATISFIABILITY PROBLEM

Let $\mathcal{X} = \left(X_j\right)_{j=0}^{n-1}$ be a set of n Boolean variables. A *literal* has the form X^δ, with $X \in \mathcal{X}$, and $\delta \in \{0, 1\}$: $X^1 = X$ and $X^0 = \neg X$. A *clause* is a disjunction of literals, and a *conjunctive form* (CF) is a conjunction of clauses. An assignment is a point $\varepsilon = \left(\varepsilon_j\right)_{j=1}^n \in \{0, 1\}^n$ in the n-dimensional hypercube. Such an assignment *satisfies* the literal X_j^δ if and only if $\varepsilon_j = \delta$; it *satisfies* a clause whenever it satisfies a literal in the clause; and it *satisfies* a CF whenever it satisfies all clauses in the CF. An *m-clause* is a clause consisting of exactly m literals, and an *m-CF* is a CF consisting just of *m*-clauses.

The *satisfiability problem* SAT consists of deciding whether a given CF has a satisfying assignment. SAT is NP-complete and 3-SAT (the restriction of SAT to 3-CF's) is also NP-complete.

For any clause C, let $h_C : \{0, 1\}^n \to \mathbb{R}$, be the map such that

$$\varepsilon \text{ satisfies } C \implies h_C(\varepsilon) = 0,$$
$$\varepsilon \text{ does not satisfy } C \implies h_C(\varepsilon) = 1.$$

And for any CF $\phi = (C_i)_{i=0}^{m-1}$ let $h_\phi : \{0, 1\}^n \to \mathbb{R}$ be $h_\phi = \sum_{i=0}^{m-1} h_{C_i}$. Clearly:

$$\forall \varepsilon \in \{0, 1\}^n : \left[h_\phi(\varepsilon) = 0 \iff \varepsilon \text{ satisfies } \phi \right],$$

thus deciding the satisfiability of ϕ is reduced to decide whether the global minimum of h_ϕ is 0.

4.1.2   AQC FORMULATION OF SAT

Let $|0\rangle = \begin{bmatrix} 1 & 0 \end{bmatrix}^T$ and $|1\rangle = \begin{bmatrix} 0 & 1 \end{bmatrix}^T$ be the vectors in the canonical basis of the Hilbert space $\mathbb{H}_1 = \mathbb{C}^2$. Let, for each $n > 1$, $\mathbb{H}_n = \mathbb{H}_{n-1} \otimes \mathbb{H}_1$ be the n-fold tensor power of $\mathbb{H}_1$. A basis of $\mathbb{H}_n$ is $(|\varepsilon\rangle)_{\varepsilon \in \{0,1\}^n}$ where

$$\varepsilon = (\varepsilon_j)_{j=1}^n \implies |\varepsilon\rangle = \bigotimes_{j=1}^n |\varepsilon_j\rangle.$$

Let $\sigma_z : \mathbb{H}_1 \to \mathbb{H}_1$ be the Pauli quantum gate with matrix $\sigma_z = \begin{bmatrix} 1 & 0 \\ 0 & -1 \end{bmatrix}$. For any bit $\delta \in \{0, 1\}$ let $\tau_{\delta z} = \frac{1}{2}(I_2 - (-1)^\delta \sigma_z)$. Independently of δ, the characteristic polynomial of $\tau_{\delta z}$ is $p_z(\lambda) = (\lambda - 1)\lambda$ and its eigenvalues are 0 and 1 with unit eigenvectors $|0\rangle$ and $|1\rangle$. The correspondence among eigenvalues and eigenvectors is determined by δ, namely:

$$\forall \varepsilon \in \{0, 1\} : \quad \tau_{\delta z} |\varepsilon\rangle = (\delta \oplus \varepsilon) |\varepsilon\rangle ; \tag{4.1}$$

in other words, if $\delta = 0$ the index of each eigenvector coincides with the eigenvalue, otherwise it is the complementary value. Thus, the zero eigenvalue of the map $\tau_{\delta z}$ corresponds to the eigenvector e_δ.

For any $\delta \in \{0, 1\}$ and $j_1 \in [\![1, n]\!]$, let $R_{E \delta j_1 n} = \bigotimes_{j_2=1}^n \rho_{z \delta j_2} : \mathbb{H}_n \to \mathbb{H}_n$ where $\rho_{z \delta j_2} = \text{identity}_{\mathbb{H}_1}$ if $j_2 \neq j_1$ and $\rho_{z \delta j_1} = \tau_{\delta z}$, thus the effect of $R_{E \delta j n}$ in an n-quregister is to apply $\tau_{\delta z}$ to the j-th qubit. Consequently,

$$\forall \varepsilon \in \{0, 1\}^n : \quad R_{E \delta j n} (|\varepsilon\rangle) = (\delta \oplus \varepsilon_j) |\varepsilon\rangle. \tag{4.2}$$

Thus, the zero eigenvalue corresponds to the basic vectors giving a satisfying assignment for the literal X_j^δ. Given a 3-clause $C = X_{j_1}^{\delta_{j_1}} \vee X_{j_2}^{\delta_{j_2}} \vee X_{j_3}^{\delta_{j_3}}$ let

$$H_{EC} = R_{E\delta_3 j_3 n} \circ R_{E\delta_2 j_2 n} \circ R_{E\delta_1 j_1 n} : \mathbb{H}_n \to \mathbb{H}_n.$$

Thus, for any $\varepsilon \in \{0, 1\}^n$, $H_{EC}(|\varepsilon\rangle) = 0$ if and only if ε satisfies the clause C; and it coincides with the linear map that on the basis vectors acts as $|\varepsilon\rangle \mapsto h_C(\varepsilon)|\varepsilon\rangle$. Thus, if $x = \sum_{\varepsilon \in \{0,1\}^n} x_\varepsilon |\varepsilon\rangle$ then $H_{EC}(x) = \sum_{\varepsilon \in \{0,1\}^n} x_\varepsilon h_C(\varepsilon)|\varepsilon\rangle$ and

$$\langle x | H_{EC}(x) \rangle = \sum_{\varepsilon \in \{0,1\}^n} \overline{x_\varepsilon} x_\varepsilon h_C(\varepsilon) = \sum_{\varepsilon \in \{0,1\}^n} |x_\varepsilon|^2 h_C(\varepsilon) \geq 0. \tag{4.3}$$

Hence, H_{EC} is a positive operator. Indeed, we have $\langle x | H_{EC}(x) \rangle = 0$ if and only if $H_{EC}(x) = 0 \in \mathbb{H}_n$, and this happens if and only if x is a linear combination of those basic vectors indexed by assignments satisfying the clause C.

For a given CF $\phi = (C_i)_{i=0}^{m-1}$ let $H_{E\phi} : \mathbb{H}_n \to \mathbb{H}_n$ be $H_{E\phi} = \sum_{i=0}^{m-1} H_{EC_i}$. Again, $H_{E\phi}$ is positive and $H_{E\phi}(x) = 0$ if and only if x is a linear combination of those basic vectors indexed by assignments satisfying the CF ϕ.

An unit n-quregister $x \in \mathbb{H}_n$ such that $H_{E\phi}(x) = 0$ is called a *ground state* for $H_{E\phi}$. Thus we have the following.

Remark 4.1   In order to find a satisfying assignment for ϕ it is sufficient to find a ground state for $H_{E\phi}$.

Let $\sigma_x : \mathbb{H}_1 \to \mathbb{H}_1$ be the Pauli quantum gate with matrix $\sigma_x = \begin{bmatrix} 0 & 1 \\ 1 & 0 \end{bmatrix}$. The map $\tau_{\delta x} = \frac{1}{2}(I_2 - (-1)^\delta \sigma_x)$ also has, independently of δ, characteristic polynomial $p_x(\lambda) = (\lambda - 1)\lambda$ and its eigenvalues are 0 and 1, now with corresponding unit eigenvectors $c_0 = \frac{1}{\sqrt{2}}(|0\rangle + |1\rangle)$ and $c_1 = \frac{1}{\sqrt{2}}(-|0\rangle + |1\rangle)$, which form an orthonormal basis of $\mathbb{H}_1$. The correspondence among eigenvalues and eigenvectors is determined as in relation (4.1) by δ, namely:

$$\forall \varepsilon \in \{0, 1\} : \quad \tau_{\delta x} c_\varepsilon = (\delta \oplus \varepsilon) c_\varepsilon. \tag{4.4}$$

Let us also make

$$\varepsilon = (\varepsilon_j)_{j=1}^n \implies c_\varepsilon = \bigotimes_{j=1}^n c_{\varepsilon_j}.$$

For any $j_1 \in [\![1, n]\!]$, let $R_{Z\delta j_1 n} = \bigotimes_{j_2=1}^n \mu_{\delta j_2} : \mathbb{H}_n \to \mathbb{H}_n$ where $\mu_{\delta j_2} = \text{identity}_{\mathbb{H}_1}$ if $j_2 \neq j_1$ and $\mu_{\delta j_1} = \tau_{\delta x}$, thus the effect of $R_{Z\delta j n}$ in an n-quregister is to apply $\tau_{\delta x}$ to the j-th qubit. Consequently, as in relation (4.2):

$$\forall \varepsilon \in \{0, 1\}^n : \quad R_{Z\delta j n}(c_\varepsilon) = (\delta \oplus \varepsilon_j) c_\varepsilon. \tag{4.5}$$

Hence, whenever $\varepsilon_j = \delta$, c_ε is a ground state of the operator $R_{Z\delta jn}$.

Let us consider $\delta = 0$ and let us write $R_{Zjn} = R_{Z0jn}$. Given a 3-clause $C = X_{j_1}^{\delta_{j_1}} \vee X_{j_2}^{\delta_{j_2}} \vee X_{j_3}^{\delta_{j_3}}$ let $H_{ZC} = R_{Zj_1n} + R_{Zj_2n} + R_{Zj_3n} : \mathbb{H}_n \to \mathbb{H}_n$. Then H_{ZC} does not depend on the "signs" $\delta_{j_1}, \delta_{j_2}, \delta_{j_3}$ of the literals, but just on the variables appearing in the clause. The following implication holds:

$$\left[\varepsilon_{j_1} = \varepsilon_{j_2} = \varepsilon_{j_3} = 0 \implies H_{ZC}(z_\varepsilon) = 0\right].$$

Given a CF $\phi = (C_i)_{i=0}^{m-1}$ let $H_{Z\phi} : \mathbb{H}_n \to \mathbb{H}_n$ be $H_{Z\phi} = \sum_{i=0}^{m-1} H_{ZC_i}$.

Remark 4.2   From relation of Equation (4.5), $c_{00\cdots0} = \frac{1}{2^{\frac{n}{2}}} \sum_{\varepsilon \in \{0,1\}^n} |\varepsilon\rangle$ is a ground state of $H_{Z\phi}$.

Remark 4.3   The following equation holds:

$$H_{Z\phi} = \sum_{j=1}^{n} d_j R_{Zjn}, \tag{4.6}$$

where, for each $j \in [\![1, n]\!]$, $d_j = \mathrm{card}\{i \in [\![1, m]\!] \mid X_j \text{ appears in } C_i\}$.

From Remark 4.2 we have that there is a "natural" ground state, $c_{00\cdots0}$, for the operator $H_{Z\phi}$, while, after Remark 4.1, to solve the SAT instance given by ϕ it is necessary to find a ground state for the operator $H_{E\phi}$. In summary, $c_{00\cdots0}$ is a ground state for $H_{Z\phi}$ but our aim is to find a ground state for $H_{E\phi}$.

For any 3-clause C, let us consider the map $I \to \mathrm{GL}(\mathbb{H}_n)$, where $\mathrm{GL}(\mathbb{H}_n)$ is the group of invertible linear automorphisms of the space $\mathbb{H}_n$, and $I = [0, 1]$ is the unit real interval, given as $t \mapsto H_C(t) = (1 - t)H_{ZC} + tH_{EC}$.

For a CF $\phi = (C_i)_{i=0}^{m-1}$, let

$$H_\phi : t \mapsto H_\phi(t) = \sum_{i=0}^{m-1} H_{C_i}(t) = \sum_{i=0}^{m-1} [(1 - t)H_{ZC} + tH_{EC}].$$

Figure 4.1 below shows an example of eigenvalue paths for a three qubits Hamiltonian H_ϕ. Let

$$\forall t \in [0, 1] : \ i\frac{d}{dt}\psi(t) = H_\phi(t)\psi(t). \tag{4.7}$$

be the proper Schrödinger equation, with Hamiltonian H_ϕ.

Let $\{\eta_v\}_{v=0}^{2^n-1} \subset (\mathbb{R}^I)^{2^n}$ be the sequence of curves giving the eigenvalues of H_ϕ (indexed according to their absolute values at the initial points for $t = 0$). Then it is possible to see that

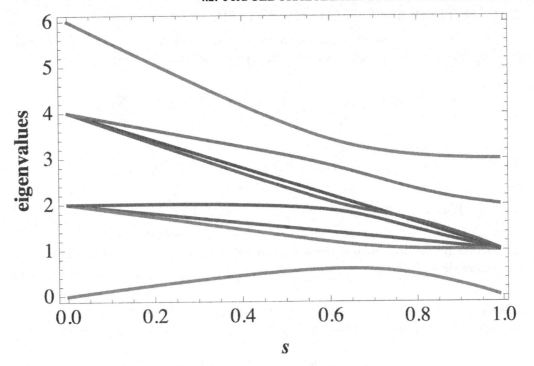

Figure 4.1: Eigenvalue paths of a three qubits Hamiltonian for a FC $\phi_3 = C_{imply}^{1,2} \wedge C_{disagree}^{1,3} \wedge C_{agree}^{2,3}$ where $C_{imply}^{1,2} = X_1^0 \vee X_2^1$, $C_{disagree}^{1,3} = (X_2^0 \vee X_3^0) \wedge (X_2^1 \vee X_3^1)$ and $C_{agree}^{2,3} = (X_1^0 \vee X_3^1) \wedge (X_1^1 \vee X_3^0)$.

η_0 and η_1 never cross on I, and, by the Adiabatic Theorem, there exists a $t_0 > 0$ such that the solutions ψ_{t_0} of the "scaled" equation

$$\forall t \in [0, t_0] : i \frac{d}{dt} \psi_{t_0}(t) = H_\phi \left(\frac{t}{t_0} \right) \psi_{t_0}(t) \tag{4.8}$$

are such that $\psi_{t_0}(t)$ gets arbitrarily close, as $t \nearrow t_0$, to a ground state for $H_{E\phi}$. A measurement of such ground state provides an assignment that either satisfies ϕ or maximizes the number of satisfied clauses in ϕ.

4.2 PROCEDURAL HAMILTONIAN CONSTRUCTION

In the following we describe a procedural construction of the Hamiltonian operators H_E and H_Z defined in Section 4.1.2.

4.2.1   HYPERPLANES IN THE HYPERCUBE

Let us enumerate the n-dimensional hypercube with indexes in $[\![0, 2^n - 1]\!]$ associating each $i \in [\![0, 2^n - 1]\!]$ with its length n big-endian base-2 representation:

$$i \leftrightarrow \operatorname{rev}((i)_2) = (\varepsilon_0, \ldots, \varepsilon_{n-1}) \in \{0, 1\}^n \quad \text{where } i = \sum_{\nu=0}^{n-1} \varepsilon_\nu 2^\nu. \tag{4.9}$$

By putting each such representation as the i-th row of a rectangular array, a $(2^n \times n)$-matrix $\mathbf{E} \in \{0, 1\}^{2^n \times n}$ is obtained. Let us denote by $\mathbf{e}_j^{(1)} \in \{0, 1\}^{2^n}$ its j-th column, $j = 0, \ldots, n - 1$. On one side, $\mathbf{e}_j^{(1)}$ can be written as the list $(0^{2^j} 1^{2^j})^{2^{n-1-j}} = \mathbf{e}_j^{(1)}$, and on the other side, it can be seen as the Boolean map that has as support the hyperplane $E_j^1 : \varepsilon_j = 1$. Let $\mathbf{e}_j^{(0)}$ be the 2^n-vector obtained from $\mathbf{e}_j^{(1)}$ by taking the complement value at each entry. Then $\mathbf{e}_j^{(0)} = (1^{2^j} 0^{2^j})^{2^{n-1-j}}$, and it represents the Boolean map with support the hyperplane $E_j^0 : \varepsilon_j = 0$. The following remark is therefore clear.

Remark 4.4   Each hyperplane E_j^δ is a $(n - 1)$-dimensional affine variety at the hypercube and its characteristic map can be written as the list

$$\mathbf{e}_j^{(\delta)} = (\bar{\delta}^{2^j} \delta^{2^j})^{2^{n-1-j}}.$$

The lists $\mathbf{e}_j^{(\delta)}$ are easily computable:
Procedure $(n - 1)$-DimensionalVarieties.
Input: $\delta \in \{0, 1\}$, $j \in [\![0, n - 1]\!]$ and $k \in [\![0, 2^n - 1]\!]$.
Output: The k-th entry of the list $\mathbf{e}_j^{(\delta)}$.

1. Let $k_0 := k \bmod (2^{n-1-j})$.

2. If $k_0 \geq 2^j$ then output δ else output $\bar{\delta}$.

Two $(n - 1)$-dimensional affine varieties are *parallel* if they are of the form E_j^0 and E_j^1, for some index $j \in [\![0, n - 1]\!]$.

Remark 4.5   The intersection of two parallel $(n - 1)$-dimensional varieties is empty, while the intersection of any two non-parallel $(n - 1)$-dimensional varieties is a $(n - 2)$-dimensional affine variety, thus the intersection of any two non-parallel $(n - 1)$-dimensional varieties has cardinality 2^{n-2}. Also, the intersection of three pairwise non-parallel $(n - 1)$-dimensional affine varieties has cardinality 2^{n-3}.

4.2.2   THE HAMILTONIAN OPERATOR H_E

For any $\delta \in \{0, 1\}$ and $j \in [\![0, n-1]\!]$, the transform $R_{E\delta jn} : \mathbb{H}_n \to \mathbb{H}_n$ defined in Section 4.1.2, being the tensor product of transforms represented by diagonal matrices with respect to the canonical basis, is represented, with respect to the basis $(|\varepsilon\rangle)_{\varepsilon \in \{0,1\}^n}$, by a diagonal matrix. Indeed:

Remark 4.6   The 2^n-length diagonal determining the diagonal matrix of $R_{E\delta jn}$ coincides with the list $\mathbf{e}_j^{(\delta)} = (\bar{\delta}^{2^j} \delta^{2^j})^{2^{n-1-j}}$.

For a 3-clause $C = X_{j_1}^{\delta_{j_1}} \vee X_{j_2}^{\delta_{j_2}} \vee X_{j_3}^{\delta_{j_3}}$, the operator $H_{EC} = R_{E\delta_3 j_3 n} \circ R_{E\delta_2 j_2 n} \circ R_{E\delta_1 j_1 n}$ is also represented by a diagonal matrix and its diagonal is the component-wise product of the lists $\mathbf{e}_{j_1}^{(\delta_1)}$, $\mathbf{e}_{j_2}^{(\delta_2)}$ and $\mathbf{e}_{j_3}^{(\delta_3)}$. Since the indexes j_1, j_2, j_3 are pairwise different, the lists are the characteristic maps of three pairwise non-parallel $(n-1)$-dimensional affine varieties. From Remark 4.5 we have the following.

Remark 4.7   With respect to the canonical basis $(|\varepsilon\rangle)_{\varepsilon \in \{0,1\}^n}$ of $\mathbb{H}_n$, for any 3-clause $C = X_{j_1}^{\delta_{j_1}} \vee X_{j_2}^{\delta_{j_2}} \vee X_{j_3}^{\delta_{j_3}}$, the operator H_{EC} is represented by a diagonal matrix and its diagonal, $D_C(C) = D_C\left((j_1, \delta_1), (j_2, \delta_2), (j_3, \delta_3)\right)$, consisting of 2^{n-3} 1's, is such that each entry can be calculated by a slight modification of the procedure $(n-1)$-DimensionalVarieties outlined above. Namely:

Procedure 3-ClauseDiagonal.
Input: A 3-clause $C = \{(j_1, \delta_1), (j_2, \delta_2), (j_3, \delta_3)\}$, and $k \in [\![0, 2^n - 1]\!]$.
Output: The k-th entry of the list D_C.

1. For $r = 1$ to 3 do

   (a) $k_{r0} := k \bmod (2^{n-1-j_r})$.
   (b) If $k_{r0} \geq 2^{j_r}$ then $x_r := \delta_r$ else $x_r := \bar{\delta}_r$.

2. Output $x_1 \cdot x_2 \cdot x_3$.

Remark 4.8   With respect to the canonical basis $(|\varepsilon\rangle)_{\varepsilon \in \{0,1\}^n}$ of $\mathbb{H}_n$, for any CF $\phi = (C_i)_{i=0}^{m-1}$ the operator $H_{E\phi} = \sum_{i=0}^{m-1} H_{EC_i}$ is represented by a diagonal matrix, and its diagonal is $D_F(\phi) = \sum_{i=0}^{m-1} D_C(C_i)$.

For any 3-clause C, let $\mathrm{Spt}_C(C) = \{j \in [\![0, 2^n - 1]\!]\mid D_C(C)[j] \neq 0\}$ be the collection of indexes corresponding to non-zero entries at the vector in the diagonal $D_C(C)$. Then $\mathrm{card}(\mathrm{Spt}_C(C)) = 2^{n-3}$. Similarly, let $\mathrm{Spt}_F(\phi)$ be the collection of indexes corresponding to non-zero entries at the vector in the diagonal $D_F(\phi)$. Clearly:

$$\phi = (C_i)_{i=0}^{m-1} \implies \mathrm{Spt}_F(\phi) = \bigcup_{i=0}^{m-1} \mathrm{Spt}_C(C_i).$$

The entries at $D_F(\phi)$ are the eigenvalues of the operator $H_{E\phi}$, and the satisfying assignments are determined by the eigenvectors corresponding to the zero eigenvalue (if zero indeed is an eigenvalue). From Remark 4.1, we obtain the following results.

Remark 4.9   Any zero entry in the 2^n-vector $D_F(\phi)$ determines a satisfying assignment for ϕ. Namely, if $D_F(\phi)[i] = 0$ then $\phi(\text{rev}\,((i)_2)) = \text{True}$.

This can also be stated as follows.

Remark 4.10   For a given CF $\phi = (C_i)_{i=0}^{m-1}$, ϕ is satisfiable if and only if the following happens $\text{Spt}_F(\phi) \neq [\![0, 2^n - 1]\!]$.

Thus, the satisfiability problem can be rephrased as follows.

Problem QASAT.

Instance: A CF $\phi = (C_i)_{i=0}^{m-1}$.

Solution: "Yes" if $\text{Spt}_F(\phi) \neq [\![0, 2^n - 1]\!]$; "No," if $\text{Spt}_F(\phi) = [\![0, 2^n - 1]\!]$.

SAT is thus reducible to QUSAT in polynomial time, consequently QUSAT is NP-complete as well.

As a second construction of the vector at the diagonal $D_C(C)$ for any 3-clause, let us enumerate these clauses in another rather conventional manner.

In a general setting, let $k \geq 3$. Then the number of k-clauses, $C = \bigvee_{j \in J} X_j^{\delta_j}$, with $\text{card}(J) = k$, in n variables, is $v_{kn} = \binom{n}{k}2^k$. For any $i \in [\![0, v_{kn} - 1]\!]$ let $i_0 = i \bmod 2^k$ and $i_1 = (i - i_0)/2^k$. Then the map $\eta : i \mapsto (i_1, i_0)$ allows us to identify $[\![0, v_{kn} - 1]\!]$ with the Cartesian product $[\![0, \binom{n}{k} - 1]\!] \times [\![0, 2^k - 1]\!]$. The map η can also be seen as the function that to each index $i \in [\![0, v_{kn} - 1]\!]$ associates the clause $C = \bigvee_{j \in J_{i_1}} X_j^{\delta_j}$ where J_{i_1} is the i_1-th k-set of $[\![0, n - 1]\!]$ and $i_0 = \sum_{\kappa=0}^{k-1} \delta_{j_\kappa} 2^\kappa$.

Remark 4.11   Let $C = X_{j_1}^{\delta_{j_1}} \vee X_{j_2}^{\delta_{j_2}} \vee X_{j_3}^{\delta_{j_3}}$ be a 3-clause, $0 \leq j_1 < j_2 < j_3 < n$. Then the collection $\text{Spt}_C(C)$ of indexes corresponding to non-zero entries at $D_C(C)$ is characterized as follows. For any $k \in [\![0, 2^n - 1]\!]$, $k \in \text{Spt}_C(C) \iff$

$$\exists (k_0, k_1, k_2, k_3) \in K :$$
$$(k_1 = \delta_1 \bmod 2)\,\&\,(k_2 = \delta_2 \bmod 2)\,\&\,(k_3 = \delta_3 \bmod 2)\,\&$$
$$k = k_0 + 2^{j_1}k_1 + 2^{j_2}k_2 + 2^{j_3}k_3,$$

where $K = [\![0, 2^{j_1} - 1]\!] \times [\![0, 2^{j_2-j_1} - 1]\!] \times [\![0, 2^{j_3-j_2} - 1]\!] \times [\![0, 2^{n-j_3} - 1]\!]$.

Remark 4.11 is consistent with the calculated cardinality of $\text{Spt}_C(C)$ because: $2^{n-3} = 2^{j_1}2^{j_2-j_1-1}2^{j_3-j_2-1}2^{n-j_3-1}$; moreover, it justifies an algorithm to compute $D_C(C)$. Namely:

Procedure 3-ClauseDiagonalBis.

Input: A 3-clause $C = \{(j_1, \delta_1), (j_2, \delta_2), (j_3, \delta_3)\}$, and $k \in [\![0, 2^n - 1]\!]$.

Output: The k-th entry of the list D_C.

1. $flg := \text{True}$; $crk := k$;

2. $k_0 := crk \bmod 2^{j_1}$; $crk := (crk - k_0)/2^{j_1}$;

3. $k_1 := crk \bmod 2^{j_2 - j_1}$; $crk := (crk - k_1)/2^{j_2 - j_1}$;

4. $flg := (k_1 == \delta_1 \bmod 2)$;

5. If flg then

    (a) $k_2 := crk \bmod 2^{j_3 - j_2}$; $crk := (crk - k_2)/2^{j_3 - j_2}$;

    (b) $flg := (k_2 == \delta_2 \bmod 2)$;

    (c) If flg then

        i. $k_3 := crk \bmod 2^{j_3 - j_2}$; $crk := (crk - k_3)/2^{n - j_3}$;

        ii. $flg := (k_3 == \delta_3 \bmod 2)$;

6. If flg then $b := 1$ else $b := 0$;

7. Output b.

4.2.3   THE HAMILTONIAN OPERATOR H_{Z_ϕ}

Now let us consider the operators with subindex Z defined in Section 4.1.2.

Let us define the following matrices:

$$
\begin{array}{llllll}
A_0 & = & [1] & ; & B_0 & = & [1] \\
A_1 & = & I_2 \otimes A_0 - \frac{1}{2}\sigma_x \otimes B_0 & ; & B_1 & = & I_2 \otimes B_0 \\
A_2 & = & I_2 \otimes A_1 - \frac{1}{2}\sigma_x \otimes B_1 & ; & B_2 & = & I_2 \otimes B_1 \\
A_3 & = & I_2 \otimes (\frac{1}{2}B_2 + A_2) - \frac{1}{2}\sigma_x \otimes B_2 & ; & B_3 & = & I_2 \otimes B_2
\end{array}
\tag{4.10}
$$

where I_2 is the (2×2)-identity matrix. For each $k \leq 3$, A_k, B_k are matrices of order $(2^k \times 2^k)$, indeed we have $B_k = I_{2^k}$.

For $n = 3$ and any 3-clause $C_{012} = X_0^{\delta_0} \vee X_1^{\delta_1} \vee X_2^{\delta_2}$ involving the three variables, the transform $H_{ZC_{012}} : \mathbb{H}_3 \to \mathbb{H}_3$ is represented, with respect to the canonical basis of $\mathbb{H}_3$, by the matrix

$$
H_{[012],3} = A_3 = \frac{1}{2}
\begin{pmatrix}
3 & -1 & -1 & 0 & -1 & 0 & 0 & 0 \\
-1 & 3 & 0 & -1 & 0 & -1 & 0 & 0 \\
-1 & 0 & 3 & -1 & 0 & 0 & -1 & 0 \\
0 & -1 & -1 & 3 & 0 & 0 & 0 & -1 \\
-1 & 0 & 0 & 0 & 3 & -1 & -1 & 0 \\
0 & -1 & 0 & 0 & -1 & 3 & 0 & -1 \\
0 & 0 & -1 & 0 & -1 & 0 & 3 & -1 \\
0 & 0 & 0 & -1 & 0 & -1 & -1 & 3
\end{pmatrix}
\tag{4.11}
$$

which is a band matrix with the following properties: its upper-right boundary is its diagonal at distance $4 = 2^{3-1}$ above the main diagonal, the lower-left boundary is also at distance 4 below the main diagonal, the main diagonal has constant value $\frac{3}{2}$, and the only values appearing in the matrix are $\frac{3}{2}, 0, -\frac{1}{2}$.

Naturally, for any $n > 3$ the transform $HZC_{012} : \mathbb{H}_n \to \mathbb{H}_n$ is represented by the matrix

$$H_{[012],n} = H_{[012],n-1} \otimes I_2. \tag{4.12}$$

The tensor product at Equation (4.12) substitutes each current entry at $H_{[012],n-1}$ by the product of that entry by the (2×2)-identity matrix. Thus, also $H_{[012],n}$ is a band matrix as its boundaries are diagonals at distance 2^{n-1} from the main diagonal, the main diagonal has constant value $\frac{3}{2}$, and the only values appearing in the matrix are $\frac{3}{2}, 0, -\frac{1}{2}$. The following algorithm results:

Procedure HZ012.

Input: An integer $n \geq 3$ and a pair $(i, j) \in [\![0, 2^n - 1]\!]^2$.

Output: The (i, j)-th entry of the matrix $H_{[012],n}$.

1. $k := j - i$;

2. Case k of

   $0 : v := \frac{3}{2}.$

   $\pm 2^\ell$: (a power of 2, with $\ell \geq 1$)

        i. $\iota := \min\{i, j\}$;

        ii. $\iota_0 := \iota \bmod 2^{\ell+1}$;

        iii. If $\iota_0 < 2^\ell$ Then $v := -\frac{1}{2}$ Else $v := 0$;

   Else: $v := 0$;

3. Output v.

For an arbitrary 3-clause $C_{j_1 j_2 j_3} = X_{j_1}^{\delta_1} \vee X_{j_2}^{\delta_2} \vee X_{j_3}^{\delta_3}$, with $0 \leq j_1 < j_2 < j_3 < n$ let $\pi_{j_1 j_2 j_3}$ be a permutation $[\![0, n - 1]\!] \to [\![0, n - 1]\!]$ such that $j_1 \mapsto 0$, $j_2 \mapsto 1$, $j_3 \mapsto 2$ and the restriction $\pi_{j_1 j_2 j_3}|_{[\![0,n-1]\!]-\{j_1,j_2,j_3\}}$ is a bijection $[\![0, n - 1]\!] - \{j_1, j_2, j_3\} \to [\![3, n - 1]\!]$. Then, there is a permutation $\rho_{j_1 j_2 j_3} : [\![0, 2^n - 1]\!] \to [\![0, 2^n - 1]\!]$, which can be determined in terms of $\pi_{j_1 j_2 j_3}$, such that the matrix $H_{[j_1 j_2 j_3],n}$ representing the transform $HZC_{j_1 j_2 j_3}$ is the action of $\rho_{j_1 j_2 j_3}$ over rows and columns on the matrix $H_{[012],n}$. Namely, let

$$\rho_{j_1 j_2 j_3} : [\![0, 2^n - 1]\!] \to [\![0, 2^n - 1]\!] \ , \ \sum_{\kappa=0}^{n-1} \varepsilon_\kappa 2^\kappa \mapsto \sum_{\kappa=0}^{n-1} \varepsilon_{\pi_{j_1 j_2 j_3}(\kappa)} 2^\kappa, \tag{4.13}$$

then when writing $H_{[012],n} = \left[h_{ij}^{(0)} \right]_{0 \leq i, j \leq 2^n - 1}$ one has

$$H_{[j_1 j_2 j_3],n} = \left[h_{\rho_{j_1 j_2 j_3}(i) \, \rho_{j_1 j_2 j_3}(j)}^{(0)} \right]_{0 \leq i, j \leq 2^n - 1}.$$

The following algorithm results:

Procedure HZFor3Clauses.

Input: An integer $n \geq 3$, a 3-clause $C = \{(j_1, \delta_1), (j_2, \delta_2), (j_3, \delta_3)\}$ and a pair $(i, j) \in [\![0, 2^n - 1]\!]^2$.

Output: The (i, j)-th entry of the matrix $H_{[j_1 j_2 j_3], n}$.

1. Compute the permutation $\rho_{j_1 j_2 j_3} : [\![0, n-1]\!] \to [\![0, n-1]\!]$ as in (4.13) ;

2. Output HZ012$[n; (\rho_{j_1 j_2 j_3}(i), \rho_{j_1 j_2 j_3}(j))]$.

(Evidently, the permutation $\rho_{j_1 j_2 j_3}$ can be computed as a preprocess to be used later for several entries (i, j).)

For a CF $\phi = (C_i = \{(j_{i1}, \delta_{i1}), (j_{i2}, \delta_{i2}), (j_{i3}, \delta_{i3}))_{i=0}^{m-1}$, the Hamiltonian operator $H_{Z\phi}$: $\mathbb{H}_n \to \mathbb{H}_n$ is represented by the matrix $H_{\phi, n} = \sum_{i=0}^{m-1} H_{[j_{i1} j_{i2} j_{i3}], n}$. Thus it can be computed directly by an iteration of algorithm HZFor3Clauses.

Remark 4.12   The ground eigenvector of the matrix $H_{\phi, n}$ will tend to a ground state of the matrix $H_{E\phi}$ when solving the Schrödinger Equation (4.8), thus providing a solution of SAT for the instance ϕ.

In this chapter we provided procedures to construct the initial and final Hamiltonians H_E and H_Z, respectively. The procedure 3-ClauseDiagonal construct the diagonal elements of the operator H_{EC} for each clause C, as the intersection of three pairwise non-parallel $(n-1)$-dimensional varieties (see Remark 4.5). The diagonal elements of $H_{E\phi}$ is the pairwise addition of the diagonal elements of the operators H_{EC} for every clause C. Although we described the diagonal elements of $H_{E\phi}$, this construction still requires an exponential number of operations, namely $m2^n$ where m is the number of clauses and n is the number of Boolean variables.

On the other hand, in Section 4.2.3 it is shown a recursive construction of the Hamiltonian H_Z that corresponds to a sparse matrix. The procedure HZFor3Clauses returns the (i, j)-th entry of the matrix $H_{[j_1 j_2 j_3], n}$ and for a 3-CF ϕ the Hamiltonian operator $H_{Z\phi}$ is constructed by iterative calls to the procedure HZFor3Clauses.

In Chapter 5 we show two constructions of the initial Hamiltonian operator for AQC, one based on the Pauli matrices and other on the Hadamard transform.

CHAPTER 5

AQC for Pseudo-Boolean Optimization

The pseudo-Boolean maps [20] appears naturally in many areas of mathematics such as in combinatorial optimization, operation research, integer programming, and artificial intelligence. Its importance is in modeling many branches of optimization problems into a general scheme based on quadratic forms. This general scheme can be used to design AQC algorithms by means of the Adiabatic Theorem in order to optimize the underlying pseudo-Boolean maps.

In this chapter, the *pseudo-Boolean maps* are introduced to build a general model for optimization in combinatorial problems. In particular the quadratic pseudo-Boolean maps are described in order to express combinatorial graph problems. We develop the Adiabatic Quantum Optimization (AQO) for pseudo-Boolean maps and we prove that the defined Hamiltonians are two-local.

We also give a general algorithm to transform any Monadic Second-Order Logic sentence into a pseudo-Boolean map, and its corresponding optimization problem for AQO. This part can be seen independently from the above results, and it can be considered as a general framework for optimization that is not restricted to graph problems.

In the background the basis for AQC given in Chapter 4 is assumed.

5.1 BASIC TRANSFORMATIONS

Let $Q = \{0, 1\}$ be the set of the integer values 0 and 1. A *Boolean function* on n variables is a function on Q^n into Q^n, where n is a positive integer and Q^n denotes the n-fold Cartesian product of Q with itself.

A *pseudo-Boolean map* of n variables is a function $f : Q^n \to \mathbb{R}$, where n is a positive integer. Consider the following problem:

Pseudo-Boolean Optimization
Instance: A pseudo-Boolean map $f : Q^n \to \mathbb{R}$.
Solution: A minimum point $x^* = \arg\min_{x \in Q^n} f(x)$.

In the following we will deal with the Pseudo-Boolean Optimization Problem.

The set of n Boolean variables will be denoted as $X = \{x_i : 0 \leq i \leq n - 1\}$ and the set of literals will be denoted as $L = \{x_i, \overline{x}_i : 0 \leq i \leq n - 1\}$ where $\overline{x}_i := 1 - x_i$.

Note that Q^n is in correspondence with the power set of $[\![0, n-1]\!]$. The following theorem asserts that any pseudo-Boolean map can be expressed as a real-valued map from $\mathcal{P}([\![0, n-1]\!])$ to $\mathbb{R}$.

Theorem 5.1 For every pseudo-Boolean map $f : Q^n \to \mathbb{R}$ on n Boolean variables, there exists a unique mapping $c : \mathcal{P}([\![0, n-1]\!]) \to \mathbb{R}$ such that

$$f(X) = \sum_{S \in \mathcal{P}([\![0,n-1]\!])} c(S) \prod_{j \in S} x_j. \tag{5.1}$$

See a proof in [20].

The size of the largest subset $S \in \mathcal{P}([\![0, n-1]\!])$ for which $c(S) \neq 0$ is called the degree of f, and is denoted by $\deg(f)$.

Remark 5.2 Every pseudo-Boolean map has a unique multilinear polynomial representation as in Equation (5.1).

A *posiform* is a polynomial expression with non-negative terms of the form:

$$\phi(L) = \sum_{k=0}^{m-1} b_k \left(\prod_{i \in A_k} x_i \right) \left(\prod_{j \in B_k} \overline{x}_j \right), \tag{5.2}$$

where $b_k \in \mathbb{R}$ and $b_k > 0$; $A_k, B_k \subseteq [\![0, n-1]\!]$, $A_k \cap B_k = \emptyset$, and $A_k \cup B_k \neq \emptyset$ for all $k = 0, \ldots, m-1$.

Proposition 5.3 Any pseudo-Boolean map can be represented as a posiform.

Proof. Let $f : Q^n \to \mathbb{R}$ be a pseudo-Boolean map defined as in (5.1). For any $S \subseteq [\![0, n-1]\!]$, let t_S be the term in f determined by S as $t_S = c(S) \prod_{j \in S} x_j$, and let π_S be a permutation of the elements in S. A posiform is an expression in which for every $S \subseteq [\![0, n-1]\!] : c(S) \geq 0$. Thus, if $c(S) < 0$ then t_S can be written as

$$t_S = c(S)(1 - \overline{x}_{\pi_S(0)} - x_{\pi_S(0)}\overline{x}_{\pi_S(1)} - \cdots - x_{\pi_S(0)} \cdots x_{\pi_S(m-2)}\overline{x}_{\pi_S(m-1)})$$

where $m = \operatorname{card} S$. Repeating this transformation for every negative term of f eventually produces a posiform of f. $\square$

From Proposition 5.3, it can be seen that any pseudo-Boolean map can have many different posiforms representing it.

Of particular interest are the quadratic pseudo-Boolean maps $f_{ue} : Q^n \to \mathbb{R}$ (i.e., $\deg(f_{ue}) \leq 2$) expressed by polynomials of the form

$$f_{ue}(X) = \sum_{j \in [\![0,n-1]\!]} u_j x_j + \sum_{\{i,j\} \in [\![0,n-1]\!]^{(2)}} e_{ij} x_i x_j, \tag{5.3}$$

for some coefficient vector $u \in \mathbb{R}^n$ and coefficient matrix $e \in \mathbb{R}^{\frac{n(n-1)}{2}}$.

For instance, given a graph $\mathbf{G} = (V, E)$, with $V = [\![0, n-1]\!]$ and $E \subseteq [\![0, n-1]\!]^{(2)}$, a representative quadratic pseudo-Boolean map is obtained as

$$f_{\mathbf{G}}(X) = \sum_{j \in [\![0,n-1]\!]} x_j - \sum_{\{i,j\} \in E} x_i x_j.$$

It can be seen that the problem to find a maximal independent vertex subset in $\mathbf{G}$ is equivalent to maximize the map $f_{\mathbf{G}}(X)$ over the hypercube Q^n.

Also, quadratic maps can be considered over the n-fold Cartesian power of the set $\mathbb{S} = \{-1, +1\}$. In fact, we have the following.

Proposition 5.4   Any maximization problem of a quadratic pseudo-Boolean map over the hypercube Q^n is equivalent to a minimization problem of a quadratic map over the power $\mathbb{S}^n$. In symbols: $\forall e \in \mathbb{R}^{\frac{n(n-1)}{2}}$, $u \in \mathbb{R}^n$ $\exists w \in \mathbb{R}^n$ $\varepsilon \in Q^n$:

$$\varepsilon = \arg\max_{Q^n} f_{we}(X) \iff \theta(\varepsilon) = \arg\min_{Q^n} f_{we}(X). \tag{5.4}$$

Proposition 5.5   Every pseudo-Boolean function f expressed as in Equation (5.1) can be reduced to a quadratic pseudo-Boolean function.

Let us consider the following algorithm:

Procedure Reduce.
Input: A pseudo-Boolean function $S_f = \sum_{S \subseteq [\![0,n-1]\!]} c_S \prod_{j \in S} X_j$.
Output: A quadratic pseudo-Boolean function f_{ue}.

1. $M = 1 + \sum_{S \subseteq [\![0,n-1]\!]} |c_S|$; $m = n$;

2. While $\exists S^* \subseteq [\![0, n-1]\!]$ with $(|S^*| > 2 \,\&\, c_{S^*} \neq 0)$ do

   Choose $\{i, j\} \subset S^*$ and let

   $c_{\{i,j\}} := c_{\{i,j\}} + M$;

   $c_{\{i,m+1\}} := -2M$; $c_{\{j,m+1\}} := -2M$;

   $c_{\{m+1\}} := 3M$;

   For all subsets $S \supseteq \{i, j\}$ with $c_S \neq 0$ define

   $c_{(S \setminus \{i,j\}) \cup \{m+1\}} := c_S$;

   $c_S := 0$;

   $m := m + 1$;

3. Output $f_{ue} := \sum_{S \subseteq [\![0,m-1]\!]} c_S \prod_{k \in S} X_k$.

Quadratic pseudo-Boolean maps appear naturally in many areas of mathematics. For instance, consider the following.

Proposition 5.6 Any instance of the 3-SAT problem can be reduced to a quadratic pseudo-Boolean map.

Proof. Let $\phi = (C_i)_{i=0}^{m-1}$ be an instance of the 3-SAT problem i.e., a 3-CF over the set of Boolean variables $\mathcal{X} = \{X_j | 0 \leq j \leq n-1\}$, where $C_i = X_{j_1}^{\delta_{j_1}} \vee X_{j_2}^{\delta_{j_2}} \vee X_{j_3}^{\delta_{j_3}}$. Let $f_\phi : Q^n \to \mathbb{R}$ be the map defined as

$$f_\phi(X) = \sum_{C_i \in \phi} (\delta_{j_1} + (-1)^{\delta_{j_1}} X_{j_1})(\delta_{j_2} + (-1)^{\delta_{j_2}} X_{j_2})(\delta_{j_3} + (-1)^{\delta_{j_3}} X_{j_3})$$

such that $f_\phi(\varepsilon) = 0$ if and only if ε satisfies ϕ, for some assignment $\varepsilon \in Q^n$. By given as input f_ϕ to the algorithm Reduce, f_ϕ can be reduced to a quadratic pseudo-Boolean map. $\square$

From Proposition 5.6 the following result is obtained.

Remark 5.7 For every 3-CF ϕ, minimizing h_ϕ as defined in Section 4.1.1 is equivalent to minimize f_ϕ.

5.2 AQC FOR QUADRATIC PSEUDO-BOOLEAN MAPS

The adiabatic quantum optimization (AQO) developed in Chapter 4 can be generalized as follows.

Let $(|\varepsilon\rangle)_{\varepsilon \in Q^n}$ be an orthonormal basis for $\mathbb{H}_n$ also-called the *computational basis*.
Given a pseudo-Boolean map $f : Q^n \to \mathbb{R}$, let us define

$$H_f : \mathbb{H}_n \to \mathbb{H}_n, \qquad H_f = \sum_{\varepsilon \in Q^n} f(\varepsilon) |\varepsilon\rangle \langle\varepsilon| . \qquad (5.5)$$

For any $\mathbf{x} \in \mathbb{H}_n$, if $\mathbf{x} = \sum_{\varepsilon \in Q^n} x_\varepsilon |\varepsilon\rangle$ then $H_f(\mathbf{x}) = \sum_{\varepsilon \in Q^n} x_\varepsilon f(\varepsilon) |\varepsilon\rangle$ and consequently

$$\langle \mathbf{x} | H_f(\mathbf{x}) \rangle = \sum_{\varepsilon \in Q^n} |x_\varepsilon|^2 f(\varepsilon). \qquad (5.6)$$

From Equation (5.6), H_f is a positive operator and $\forall \varepsilon \in Q^n : H_f |\varepsilon\rangle = f(\varepsilon) |\varepsilon\rangle$, then H_f is diagonal in the computational basis, i.e., $\operatorname{diag}(H_f) = (f(\varepsilon))_{\varepsilon \in Q^n}$.

In order to construct explicitly H_f, it is necessary to evaluate the map f at every point in Q^n (see Chapter 4).

The Hamiltonian problem H_f for AQO is constructed as a sum of one-dimensional projectors along every possible direction in the computational basis. Now, let us define a more convenient Hamiltonian for quadratic pseudo-Boolean maps.

Let $f_{ue} : Q^n \rightarrow \mathbb{R}$ be a quadratic pseudo-Boolean map of the form:

$$f_{ue}(X) = a + \sum_{j \in A} u_j x_j + \sum_{\{i,j\} \in B} e_{ij} x_i x_j, \tag{5.7}$$

where $A \subseteq [\![0, n-1]\!]$, $B \subseteq [\![0, n-1]\!]^{(2)}$, and $\forall j \in A : u_j \in \mathbb{R}$, $\forall \{i, j\} \in B : e_{ij} \in \mathbb{R}$ and $a \in \mathbb{R}$.

Remark 5.8   Any quadratic pseudo-Boolean map can be represented as in Equation (5.7).

Remark 5.8 also asserts that any quadratic pseudo-Boolean map has an inherent graph structure.

For any $j \in [\![0, n-1]\!]$ and $\delta \in \{0, 1\}$ let $\sigma_{z,\delta}^j = \bigotimes_{v=0}^{n-1} s_v : \mathbb{H}_n \rightarrow \mathbb{H}_n$ where $s_v = \frac{1}{2}(I_2 + (-1)^{\delta} \sigma_z)$ if $v = j$ and $s_v = \mathrm{Id}$ otherwise.

Let $H_{f_{ue}} : \mathbb{H}_n \rightarrow \mathbb{H}_n$ be defined as follows:

$$H_{f_{ue}} = a(\sigma_{z,0}^b + \sigma_{z,1}^b) + \sum_{j \in A} u_j \sigma_{z,0}^j + \sum_{\{i,j\} \in B} e_{ij} \sigma_{z,0}^i \sigma_{z,0}^j, \tag{5.8}$$

where $b \in [\![0, n-1]\!]$.

For any $\varepsilon \in Q^n : H_{f_{ue}} |\varepsilon\rangle = f_{ue}(\varepsilon) |\varepsilon\rangle$, then $H_{f_{ue}}$ is diagonal in the computational basis.

Similar constructions for some specific combinatorial graph problems can be seen in [24–26, 75, 76].

The construction of $H_{f_{ue}}$ is more efficient than the construction of H_f. H_f is an addition of 2^n projections, while $H_{f_{ue}}$ is a sum of $\mathrm{card}(A) + \mathrm{card}(B) + 1$ Pauli operator products.

The Hamiltonian problem $H_{f_{ue}}$ can be implemented by the well known Ising model in QM (see [11, 51]).

In the following we consider the construction and structure of two initial Hamiltonians for AQO, the first one is based on the Hadamard transform and the last one is based on the σ_x Pauli operator.

5.2.1   HADAMARD TRANSFORM

In the following we describe the initial Hamiltonian for AQO based on the Hadamard transform similar to the proposed in [92], and we show its matrix structure, i.e., its construction from the computational point of view.

Let us recall that the Hadamard transform is the unitary map $W : \mathbb{H}_1 \rightarrow \mathbb{H}_1$ whose matrix, relative to the canonical basis is

$$W = \frac{1}{\sqrt{2}} \begin{bmatrix} 1 & 1 \\ 1 & -1 \end{bmatrix}.$$

Let $W^{\otimes n} : \mathbb{H}_n \rightarrow \mathbb{H}_n$ be the n-fold tensor product of W. For $n \geq 1$,

$$W^{\otimes n} = (w_{ijn})_{0 \leq i, j \leq 2^n - 1} \tag{5.9}$$

such that $\forall i, j \in [\![0, 2^n - 1]\!] : w_{ijn} = \frac{1}{2^{\frac{n}{2}}}(-1)^{i \cdot j}$ where $i \cdot j$ is the bitwise dot product of the binary representations of the numbers i and j.

Remark 5.9   The set of states $(W^{\otimes n} |\varepsilon\rangle)_{\varepsilon \in Q^n}$ form an orthonormal basis for $\mathbb{H}_n$, also called the Hadamard basis.

Thus, any vector $\mathbf{x} \in \mathbb{H}_n$ can be written as $\mathbf{x} = \sum_{\varepsilon \in Q^n} x_\varepsilon W^{\otimes n} |\varepsilon\rangle$ where $\forall \varepsilon \in Q^n : x_\varepsilon \in \mathbb{C}$. Observe that $(W^{\otimes n} |\varepsilon\rangle)^H \mathbf{x} = x_\varepsilon$ for all $\varepsilon \in Q^n$ and therefore

$$\left(\sum_{\varepsilon \in Q^n} (W^{\otimes n} |\varepsilon\rangle)(W^{\otimes n} |\varepsilon\rangle)^H \right) \mathbf{x} = \sum_{\varepsilon \in Q^n} (W^{\otimes n} |\varepsilon\rangle)(W^{\otimes n} |\varepsilon\rangle)^H \mathbf{x}$$
$$= \sum_{\varepsilon \in Q^n} x_\varepsilon W^{\otimes n} |\varepsilon\rangle$$
$$= \mathbf{x}.$$

It follows that

$$\sum_{\varepsilon \in Q^n} (W^{\otimes n} |\varepsilon\rangle)(W^{\otimes n} |\varepsilon\rangle)^H = I; \tag{5.10}$$

this equation is known as the *completeness relation* [70].

Let $h : Q^n \to \mathbb{R}^+$ be a map such that $h(0^n) = 0$ and $h(\varepsilon) \geq 1$ for all $\varepsilon \in Q^n - \{0^n\}$. Let $H_h : \mathbb{H}_n \to \mathbb{H}_n$ be defined as follows:

$$H_h = W^{\otimes n} \left(\sum_{\varepsilon \in Q^n} h(\varepsilon) |\varepsilon\rangle \langle\varepsilon| \right) (W^{\otimes n})^H$$
$$= \sum_{\varepsilon \in Q^n} h(\varepsilon) (W^{\otimes n} |\varepsilon\rangle) (W^{\otimes n} |\varepsilon\rangle)^H. \tag{5.11}$$

The ground state of H_h is given by $x_0 = W^{\otimes n} |0^n\rangle = \frac{1}{2^{\frac{n}{2}}} \sum_{\varepsilon \in Q^n} |\varepsilon\rangle$. Then,

$$H_h(x_0) = \sum_{\varepsilon \in Q^n} h(\varepsilon) (W^{\otimes n} |\varepsilon\rangle) (W^{\otimes n} |\varepsilon\rangle)^H x_0 = 0,$$

i.e., x_0 is an eigenvector corresponding to the eigenvalue 0.

Now, let us describe explicitly the construction of the matrix H_h. Consider the correspondence from Q^n to $[\![0, 2^n - 1]\!]$ by the map $\varepsilon \mapsto \sum_{v=0}^{n-1} \varepsilon_v 2^v$. For each $j \in [\![0, 2^n - 1]\!]$, $W^{\otimes n} |j\rangle$ corresponds to the j-th column of $W^{\otimes n}$, and $\forall i, j \in [\![0, 2^n - 1]\!]$:

$$\left(W^{\otimes n} |i\rangle \right) \left(W^{\otimes n} |j\rangle \right)^H = (v_{kl})_{0 \leq k, l \leq 2^n - 1} \tag{5.12}$$

such that $v_{kl} = \frac{1}{2^{\frac{n}{2}}}(-1)^{k \cdot i + l \cdot j}$ and if $i = j$ then $v_{kl} = \frac{1}{2^{\frac{n}{2}}}(-1)^{(k \oplus l) \cdot i}$.

H_h can be rewritten as follows:

$$H_h = (u_{kl})_{0 \le k, l \le 2^n - 1}$$

such that

$$
\begin{aligned}
u_{kl} &= \frac{1}{2^n} \sum_{i=0}^{2^n - 1} h(i)(-1)^{(k \oplus l) \cdot i} \\
&= \frac{1}{2^n} \left[(-1)^{(k \oplus l) \cdot 0} \quad \cdots \quad (-1)^{(k \oplus l) \cdot (2^n - 1)} \right] \left[h(0) \quad \cdots \quad h(2^n - 1) \right]^T, \quad (5.13)
\end{aligned}
$$

and since $k \oplus l = l \oplus k$, then H_h is a symmetric matrix.

For every $j \in [\![0, 2^n - 1]\!]$, let $W_j^{\otimes n} = (w_{klj})_{0 \le k, l \le 2^n - 1}$ such that $w_{klj} = 2^{-\frac{n}{2}} (-1)^{(k \oplus j) \cdot l}$. From Equation (5.13), every column of H_h can be expressed as $2^{-\frac{n}{2}} W_j^{\otimes n} \mathbf{h}$ where $\mathbf{h} = \left[h(0) \quad \cdots \quad h(2^n - 1) \right]^T$.

Proposition 5.10   For every $j \in [\![0, 2^n - 1]\!]$: $P_j = W_j^{\otimes n} W^{\otimes n}$ is a permutation matrix.

Proof. A permutation matrix is a square matrix which has exactly one 1 in every row and every column, and the other elements are zeros. Now, by definition P_j is a square matrix, and let $W_j^{\otimes n} W^{\otimes n} = (w_{pq})_{0 \le p, q \le 2^n - 1}$ where $w_{pq} = \frac{1}{2^n} \sum_{l=0}^{2^n - 1} (-1)^{(j \oplus p \oplus q) \cdot l}$, and $w_{pq} = 1$ if $(j \oplus p \oplus q) = 0$ and $w_{pq} = 0$ otherwise. For fixed j, $p \in [\![0, 2^n - 1]\!]$, there is a unique $q \in [\![0, 2^n - 1]\!]$ such that $j \oplus p \oplus q = 0$, then the proposition follows. $\square$

From Proposition 5.10 the following is obtained.

Remark 5.11   For every $j \in [\![0, 2^n - 1]\!]$, the j-th column of H_h can be expressed as $2^{-\frac{n}{2}} P_j W^{\otimes n} \mathbf{h}$.

Thus, from Remark 5.11 every column of H_h can be constructed by permuting the elements of the column vector $2^{-\frac{n}{2}} W^{\otimes n} \mathbf{h}$.

5.2.2 σ_x TRANSFORM

Let us consider the Pauli transform $\sigma_x : \mathbb{H}_1 \to \mathbb{H}_1$ whose matrix with respect to the canonical basis is

$$\sigma_x = \begin{bmatrix} 0 & 1 \\ 1 & 0 \end{bmatrix}. \quad (5.14)$$

σ_x has eigenvalues $+1, -1$ with respective eigenvectors $c_0 = W \lvert 0 \rangle$ and $c_1 = W \lvert 1 \rangle$.

For every $\varepsilon \in Q^n$, let

$$c_\varepsilon = \bigotimes_{j=0}^{n-1} c_{\varepsilon_j},$$

$(c_\varepsilon)_{\varepsilon \in Q^n}$ is the Hadamard basis of $\mathbb{H}_n$.

For any index $j \in [\![0, n-1]\!]$ and $\delta \in \{0, 1\}$ let $\sigma_{x,\delta,j} = \bigotimes_{\nu=0}^{n-1} \tau_{\nu,\delta} : \mathbb{H}_n \to \mathbb{H}_n$, where $\tau_{\nu,\delta} = \frac{1}{2}(I + (-1)^\delta \sigma_x)$ if $\nu = j$ and $\tau_{\nu,\delta} = \mathrm{Id}$ otherwise. Notice that $\tau_{\nu,\delta}$ can be written as $\tau_{\nu,\delta} = c_\delta c_\delta^H$.

For any $\varepsilon \in Q^n$, $\delta \in \{0, 1\}$ and $j \in [\![0, n-1]\!]$:

$$
\begin{aligned}
\sigma_{x,\delta,j}(c_\varepsilon) &= (I \otimes \cdots \otimes c_\delta c_\delta^H \otimes \cdots \otimes I)(c_{\varepsilon_0} \otimes \cdots \otimes c_{\varepsilon_j} \otimes \cdots \otimes c_{\varepsilon_{n-1}}) \\
&= c_{\varepsilon_0} \otimes \cdots \otimes c_\delta c_\delta^H c_{\varepsilon_j} \otimes \cdots \otimes c_{\varepsilon_{n-1}} \\
&= c_{\varepsilon_0} \otimes \cdots \otimes (c_\delta, c_{\varepsilon_j}) c_\delta \otimes \cdots \otimes c_{\varepsilon_{n-1}},
\end{aligned} \tag{5.15}
$$

where $(c_\delta, c_{\varepsilon_j})$ is the inner product of c_δ and c_{ε_j}. Since that $\{c_0, c_1\}$ form a basis for $\mathbb{H}_1$, $(c_\delta, c_{\varepsilon_j}) = 1$ if $\delta = \varepsilon_j$ and $(c_\delta, c_{\varepsilon_j}) = 0$ otherwise.

From Equation (5.15) it is satisfied that

$$
\sigma_{x,\delta,j}(c_\varepsilon) = \neg(\delta \oplus \varepsilon_j) c_\varepsilon. \tag{5.16}
$$

Now, let $\Delta : [\![0, n-1]\!] \to \mathbb{R}$ be a weighting map. For any $\delta \in \{0, 1\}$ let us define the operator:

$$
H_x : \mathbb{H}_n \to \mathbb{H}_n \quad, \quad H_x = \sum_{j=0}^{n-1} \Delta(j) \sigma_{x,\delta,j}. \tag{5.17}
$$

From (5.16) it is satisfied that:

$$
\forall \varepsilon \in Q^n : \quad H_x(c_\varepsilon) = \left(\sum_{j=0}^{n-1} \neg(\delta \oplus \varepsilon_j) \Delta(j) \right) c_\varepsilon. \tag{5.18}
$$

The ground state of H_x is the state $x_0 = \frac{1}{2^{\frac{n}{2}}} \sum_{\varepsilon \in Q^n} |\varepsilon\rangle$ with corresponding eigenvalue equal to 0.

The Hamiltonians in Equations (5.11) and (5.17) are both diagonal in the Hadamard basis and can be written as follows:

$$
H_h = W^{\otimes n} D_h W^{\otimes n}, \quad H_x = W^{\otimes n} D_x W^{\otimes n},
$$

where D_h and D_x are diagonal matrices.

From Equations (5.11) and (5.18), $\mathrm{diag}(D_h) = (h(\varepsilon))_{\varepsilon \in Q^n}$ and $\mathrm{diag}(D_x) = (\eta(\varepsilon))_{\varepsilon \in Q^n}$ such that for any $\delta \in \{0, 1\}$, $\forall \varepsilon \in Q^n : \eta(\varepsilon) = \sum_{j=0}^{n-1} \neg(\delta \oplus \varepsilon_j) \Delta(j)$.

Remark 5.12 The Hamiltonian given in Equation (5.17) can be expressed as in Equation (5.11). The converse is not always true.

5.3 k-LOCAL HAMILTONIAN PROBLEMS

For each $n \in \mathbb{N}$ let Q^n be the set of n-length words over Q and let $Q^* = \bigcup_{n \geq 0} Q^n$ be the *dictionary*, i.e., the set of finite length words, of Q.

A *promise problem* consists of a partition $\{Y, N\}$ of Q^*. For any word instance $\sigma \in Q^*$ the corresponding solution is a decision whether $\sigma \in Y$ (σ is a *Yes-instance*) or $\sigma \in N$ (σ is a *No-instance*).

For each $n \in \mathbb{N}$, $B_n = (|\sigma\rangle)_{\sigma \in Q^n} \subset S_n$ is the canonical basis of $\mathbb{H}_n$. Let $B_* = \bigcup_{n \geq 0} B_n$.

A *verifier* is a map of the form $V : B_* \times B_* \to Q$. If $V(|\sigma\rangle, |\tau\rangle) = 1$ then it is said that the verifier *accepts* σ as a Yes-instance with *proof*, or *certificate*, τ.

Let $\varepsilon : \mathbb{N} \to [0, 1]$ be such that

$$\forall \sigma \in Q^* : \quad 2^{-\Omega(|\sigma|)} \leq \varepsilon(|\sigma|) \leq \frac{1}{3}. \tag{5.19}$$

The class QMA_ε consists of those promise problems $\{Y, N\}$ such that there is a quantum polynomial time verifier V satisfying:

- $\forall \sigma \in Y \; \exists \tau \in Q^* : \Pr(V(|\sigma\rangle, |\tau\rangle) = 1) \geq 1 - \varepsilon(|\sigma|)$.

- $\forall \sigma \in N \; \forall \tau \in Q^* : \Pr(V(|\sigma\rangle, |\tau\rangle) = 1) \leq \varepsilon(|\sigma|)$.

Remark 5.13 [58] If $\varepsilon_0, \varepsilon_1 : \mathbb{N} \to [0, 1]$ satisfy condition (5.19) then $\mathrm{QMA}_{\varepsilon_0} = \mathrm{QMA}_{\varepsilon_1}$.

The common class resulting from remark 5.13 is QMA. Figure 5.1 shows the contention of QMA with respect to the classes BQP, NP, and P.

Let $k \leq n$. Let $K \subset [\![0, n-1]\!]$ be an index set of cardinality k. Let

$$B_K = \{ \bigotimes_{j=0}^{n-1} \mathbf{b}_j \mid \mathbf{b}_j = |s_j\rangle \text{ with } s_j \in Q \text{ if } j \in K, \; \mathbf{b}_j = |0\rangle \text{ otherwise} \}$$

be the collection of basic vectors in $\mathbb{H}_n$ whose "non-horizontal tensor factors" appear just at indexes in K. Let $V_K = \mathcal{L}(B_K)$ be the space spanned by B_K. Then V_K is isomorphic to $\mathbb{H}_k$ and there is a complementary space V_K' isomorphic to $\mathbb{H}_{n-k}$ such that $\mathbb{H}_n = V_K \otimes V_K'$.

Let $H : \mathbb{H}_n \to \mathbb{H}_n$ be a Hamiltonian operator. It is said that H *acts on k-qubits* if there is an index set $K \subset [\![0, n-1]\!]$ of cardinality k such that there is a Hamiltonian operator $H_K : V_K \to V_K$ with

$$H = H_K \otimes \mathrm{Id}_{2^{n-k}}, \tag{5.20}$$

where $\mathrm{Id}_{2^{n-k}}$ is the identity map in the complementary space V_K'.

A Hamiltonian $H : \mathbb{H}_n \to \mathbb{H}_n$ is k-*local* if it can be expressed as the addition of Hamiltonian operators, each acting on k-qubits, with the additional conditions stated below:

1. $H = \sum_{j \in J} H_j$, with $\mathrm{card}(J) = n^{O(1)}$.

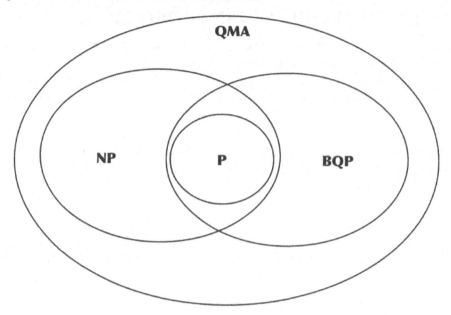

Figure 5.1: Contention of the class QMA with respect to classes NP, BQP, and P.

2. $\forall j \in J: \|H_j\| \leq n^{O(1)}$.

The second condition is equivalent to state that $\forall j$, both H_j and $\mathrm{Id}_n \quad H_j$ are non-negative.

For any Hamiltonian $H : \mathbb{H}_n \rightarrow \mathbb{H}_n$ let us denote by $\lambda_0(H)$ the smallest, in absolute value, eigenvalue of H.

Instance: k-local Hamiltonian
Solution: A k-local Hamiltonian $H : \mathbb{H}_n \rightarrow \mathbb{H}_n$ and two real numbers a, b such that $b \quad a \geq n^{\ O(1)}$ and either $\lambda_0(H) \leq a$ or $\lambda_0(H) \geq b$. 1 if H has an eigenvalue below a and 0 if all eigenvalues of H are at least b.

k-*local Hamiltonian* is NP-hard for $k \geq 2$ [95]. 5-*local Hamiltonian*, 3-*local Hamiltonian*, and 2-*local Hamiltonian* were proved QMA-complete, respectively, in [58], [57], and [56].

5.3.1   REDUCTION OF GRAPH PROBLEMS TO THE 2-LOCAL HAMILTONIAN PROBLEM

Let $\mathbf{G} = ([0, n \quad 1], E)$ be a graph. Let us assume that there is a quadratic Boolean map $f : Q^n \rightarrow \mathbb{R}$,

$$\varepsilon \mapsto f(\varepsilon) = \sum_{ij \in E} \left[a_3 \varepsilon_i \varepsilon_j + a_2(1 \quad \varepsilon_i)\varepsilon_j + a_1 \varepsilon_i(1 \quad \varepsilon_j) + a_0(1 \quad \varepsilon_i)(1 \quad \varepsilon_j) \right] \qquad (5.21)$$

that should be minimized. The map f is determined by the edges at the graph $\mathbf{G}$. Equivalently, the map f can be expressed as

$$\forall \varepsilon \in Q^n : f(\varepsilon) = \sum_{ij \in E} \left[c_0 + c_1 \varepsilon_i + c_2 \varepsilon_j + c_3 \varepsilon_i \varepsilon_j \right] \tag{5.22}$$

with

$$\begin{bmatrix} a_0 \\ a_1 \\ a_2 \\ a_3 \end{bmatrix} = \begin{bmatrix} 1 & 0 & 0 & 0 \\ 1 & 1 & 0 & 0 \\ 1 & 0 & 1 & 0 \\ 1 & 1 & 1 & 1 \end{bmatrix} \begin{bmatrix} c_0 \\ c_1 \\ c_2 \\ c_3 \end{bmatrix}. \tag{5.23}$$

Let us consider the projection

$$\pi_2 = a_0 |00\rangle \langle 00| + a_1 |01\rangle \langle 01| + a_2 |10\rangle \langle 10| + a_3 |11\rangle \langle 11| : \mathbb{H}_2 \to \mathbb{H}_2, \tag{5.24}$$

represented by the matrix $\pi_2 = \mathrm{diag}[a_0 \ a_1 \ a_2 \ a_3]$. Hence, its eigenvalues are a_0, a_1, a_2, a_3 with corresponding eigenspaces $\mathcal{L}(|00\rangle), \mathcal{L}(|01\rangle), \mathcal{L}(|10\rangle), \mathcal{L}(|11\rangle)$, respectively. Namely, for the basis vector $|\varepsilon_0 \varepsilon_1\rangle \in B_2$ we have

$$\pi_2 |\varepsilon_0 \varepsilon_1\rangle = a_{(\varepsilon_0 \varepsilon_1)_2} |\varepsilon_0 \varepsilon_1\rangle. \tag{5.25}$$

For any index pair ij let $\pi_2[ij]$ be the Hamiltonian defined by (5.20), with $K = \{i, j\}$ and $H_K = \pi_2$. The eigenvalues of each $\pi_2[ij]$ are the coefficients a_i and after (5.25)

$$\forall \varepsilon = (\varepsilon_0, \ldots, \varepsilon_{n-1}) \in Q^n : \pi_2[ij]\varepsilon = a_{(\varepsilon_i \varepsilon_j)_2} \varepsilon. \tag{5.26}$$

Let

$$H = \sum_{ij \in E} \pi_2[ij]. \tag{5.27}$$

Clearly, $\|H\| = \max_i |a_i|$, thus H is a 2-local Hamiltonian if $|a_i| \leq 1$. From (5.26)

$$\forall \varepsilon = (\varepsilon_0, \ldots, \varepsilon_{n-1}) \in Q^n : H\varepsilon = \left(\sum_{ij \in E} a_{(\varepsilon_i \varepsilon_j)_2} \right) \varepsilon. \tag{5.28}$$

Thus, the ground states correspond to the eigenvalue

$$\lambda_0(H) = \min_\varepsilon \left| \sum_{ij \in E} a_{(\varepsilon_i \varepsilon_j)_2} \right|. \tag{5.29}$$

2-local Hamiltonian would then provide a solution of the minimization problem of the quadratic Boolean map.

The above technique is just a generalization of the reduction of *Max Cut* and *Independent Set* to 2-local Hamiltonian presented in [95].

A *cut* in a weighted graph $\mathbf{G} = (\llbracket 0, n-1 \rrbracket, E, w)$, where $w : E \to \mathbb{R}^+$ is a *weighting map*, is a partition $C = \{V_0, V_1\}$ of the vertex set $\llbracket 0, n-1 \rrbracket$. For $\delta, \varepsilon \in Q$ let $E_{\delta\varepsilon} = \{ij \in E \mid i \in V_\delta, j \in V_\varepsilon\}$ be the collection of edges with an extreme in V_δ and the other in V_ε. The *weight of the cut* is $w(C) = \sum_{\delta \neq \varepsilon, ij \in E_{\delta\varepsilon}} w(ij)$.

Max Cut

Instance: A weighted graph $\mathbf{G} = (\llbracket 0, n-1 \rrbracket, E, w)$ and a threshold $w_0 \in \mathbb{R}^+$.

Solution: A decision about whether there exists a cut C such that $w(C) \geq w_0$.

Simple Max Cut is the restriction of *Max Cut* to weighted graphs with constant weights 1. Both *Max Cut* and *Simple Max Cut* are NP-complete.

Let us consider $\mathbf{G} = (\llbracket 0, n-1 \rrbracket, E, 1)$ as a weighted graph with unitary weights. Let $X = (X_j)_{j=0}^{n-1}$ be a collection of Boolean variables. For any assignment $\varepsilon = (\varepsilon_j)_{j=0}^{n-1} \in Q^n$ and each $\delta \in Q$, let $V_\delta = \{j \mid \varepsilon_j = \delta\}$. This determines a correspondence among assignments and cuts. For any cut $C = \{V_0, V_1\}$ we have

$$w(C) = \text{card}(E_{01}) + \text{card}(E_{10}) = \sum_{ij \in E} \left[(1 - \varepsilon_i)\varepsilon_j + \varepsilon_i(1 - \varepsilon_j) \right].$$

Thus, given a threshold $w_0 \in \mathbb{Z}^+$ we have

$$w(C) \geq w_0 \iff \sum_{ij \in E} \left[\varepsilon_i \varepsilon_j + (1 - \varepsilon_i)(1 - \varepsilon_j) \right] \leq \text{card}(E) - w_0.$$

Hence, *Max Cut* can be stated as an optimization problem for the quadratic Boolean map

$$Q^n \to \mathbb{R}, \ \varepsilon \mapsto \sum_{ij \in E} \left[\varepsilon_i \varepsilon_j + (1 - \varepsilon_i)(1 - \varepsilon_j) \right],$$

which is indeed of the form (5.21) with $a_0 = a_3 = 1$ and $a_1 = a_2 = 0$.

On the other side, a vertex set $V \subset \llbracket 0, n-1 \rrbracket$ is *independent* if $E[V] = \emptyset$, i.e., no edge exists among two points in V. The following is a well-known NP-complete problem.

Independent Set

Instance: A graph $\mathbf{G} = (\llbracket 0, n-1 \rrbracket, E)$ and a threshold $w_0 \in \mathbb{Z}^+$.

Solution: A decision about whether there exists an independent set V such that $\text{card}(V) \geq w_0$.

For any instance graph let us consider the quadratic Boolean map

$$f : Q^n \to \mathbb{R}, \ \varepsilon \mapsto f(\varepsilon) = \sum_{i=0}^{n-1} \varepsilon_i - \sum_{ij \in E} \varepsilon_i \varepsilon_j.$$

Thus, given a threshold $w_0 \in \mathbb{Z}^+$ we have

$$f(\varepsilon) \geq w_0 \iff g(\varepsilon) = n - f(\varepsilon) = \sum_{i=0}^{n-1} [1 - \varepsilon_i] + \sum_{ij \in E} \varepsilon_i \varepsilon_j. \leq n - w_0.$$

The second sum is a quadratic of the form (5.21) with $a_0 = a_1 = a_2 = 0$ and $a_3 = 1$. For the first sum, let us consider the projections $\rho_1 = |0\rangle \langle 0| : \mathbb{H}_1 \to \mathbb{H}_1$ and for the second $\pi_2 = |11\rangle \langle 11| :$ $\mathbb{H}_2 \to \mathbb{H}_2$. Let

$$H = \sum_{i=0}^{n-1} \rho_1[i] + \sum_{ij \in E} \pi_2[ij]. \tag{5.30}$$

By choosing $a = n - w_0 + \frac{1}{2}$ and $b = a + \frac{1}{4}$ a solution of 2-*local Hamiltonian* gives a solution of *Independent Set*.

5.4 GRAPH STRUCTURES AND OPTIMIZATION PROBLEMS

In *Descriptive Complexity* the class of problems defined in *Computational Complexity* are characterized by expressions in first- and second-order logic. There are many results in this area such as the Fagin's Theorem that asserts that the class of problems NP is equal to the set of Boolean queries in existential second-order logic [50, 63].

On the other hand, Monadic second-order logic has played an important role in expressing many NP-complete problems, for instance the graph coloring problem. Courcelle [28] showed that any problem expressible in monadic second-order logic can be stated as an optimization problem [27, 44, 50, 73].

Monadic second-order logic expressions provide a syntactical representation of many optimization problems. This representation is in correspondence with a Boolean algebra. Following [20], we consider the correspondence between second-order logic expressions and Boolean formulas, in order to define a corresponding pseudo-Boolean map. This scheme can be applied to any optimization problem expressible in monadic second-order logic, and subsequently to be solved by AQO.

The necessary terminology and definitions on first- and second-order logic are given.

5.4.1 RELATIONAL SIGNATURES

A *signature* $\Sigma = (\Phi, \Pi)$ consists of a set Φ of function symbols and positive integers $(\rho(f))_{f \in \Phi}$, and of a set of relation symbols Π and positive integers $(\rho(R))_{R \in \Pi}$. The numbers $\rho(f)$ and $\rho(R)$ assert that f is a function of $\rho(f)$ variables and R is a $\rho(R)$-ary relation.

A signature without relation symbols is called an *algebraic signature* and a signature without function symbols is called a *relational signature*.

Let $\mathcal{R}$ be a relational signature, $\mathcal{R}_0 := \{R \in \mathcal{R}|\rho(R) = 0\}$ be the set of relation symbols of arity zero called *constant symbols*, $\mathcal{R}_i := \{R \in \mathcal{R}|\rho(R) = i\}$ be the set of relation symbols of arity i, and $\mathcal{R}_+ := \bigcup\{\mathcal{R}_i|i \geq 1\}$ be the set of relation symbols of positive arity.

A $\mathcal{R}$-*structure* is a tuple $S = \langle D_S, (R_S)_{R\in\mathcal{R}_+}, (c_S)_{c\in\mathcal{R}_0}\rangle$ where D_S is a finite set called the *domain* of S, for each $R \in \mathcal{R}_+$, $R_S \subseteq D_S^{\rho(R)}$ is called the *interpretation* of R, and for each $c \in \mathcal{R}_0, c_S \in D_S$ is called the *interpretation* of c.

5.4.2 FIRST-ORDER LOGIC

Let $\mathcal{V}_0$ be a countable set of variables and let $\mathcal{R}$ be a relational signature. A *term* is either a variable in $\mathcal{V}_0$ or a constant symbol in $\mathcal{R}_0$. An *atomic formula s* is either $s = t$ or $s = R(t_1, \ldots, t_{\rho(R)})$ where $R \in \mathcal{R}_+$ and $t, t_1, \ldots, t_{\rho(R)}$ are terms. If ϕ and ψ are atomic formulas, then ϕ, $\neg\phi$, $(\phi \wedge \psi)$, $(\phi \vee \psi)$, $(\phi \Rightarrow \psi)$ and $(\phi \Leftrightarrow \psi)$ are *first-order formulas* over $\mathcal{R}$. If φ is a first-order formula and $x \in \mathcal{V}_0$, then $\exists x\varphi$ and $\forall x\varphi$ are first-order formulas as well.

It is said that a variable $x \in \mathcal{V}_0$ is *free* in a formula ϕ, if it is not inside a $\exists x$ or $\forall x$ quantifier; otherwise, it is *bound*. A formula without free variables over a relational signature $\mathcal{R}$ is called *closed*.

The set of all first-order formulas over a relational signature $\mathcal{R}$ with free variables in $\mathcal{X} \subseteq \mathcal{V}_0$ is denoted as $\text{FO}(\mathcal{R}, \mathcal{X})$. A formula $\varphi \in \text{FO}(\mathcal{R}, \{x_1, \ldots, x_n\})$ will be written as $\varphi(x_1, \ldots, x_n)$ to specify its free variables. The set of all $\mathcal{R}$-structures will be denoted as $\text{STR}(\mathcal{R})$. For a $S \in \text{STR}(\mathcal{R})$ given as $S = \langle D_S, (R_S)_{R\in\mathcal{R}_+}, (c_S)_{c\in\mathcal{R}_0}\rangle$, if $\varphi(x_1, \ldots, x_n) \in \text{FO}(\mathcal{R}, \{x_1, \ldots, x_n\})$ and $d_1, \ldots, d_n \in D_S$, then $S \models \varphi(d_1, \ldots, d_n)$ denotes that φ is true in S when $x_i = d_i$ for $i = 1, \ldots, n$.

5.4.3 SECOND-ORDER LOGIC

Let $\mathcal{V}_\omega$ be a countable set consisting of first-order variables and relation variables denoted by uppercase letters $X_1, \ldots, X_m$. Each relation variable $X \in \mathcal{V}_\omega$ has arity $\rho(X)$, and there are countably many relation variables of each arity.

The *second-order formulas* over a relational signature $\mathcal{R}$ are defined as follows. For any $R \in \mathcal{R}$, $R(t_1, \ldots, t_{\rho(R)})$ is an atomic formula where $t_1, \ldots, t_{\rho(R)}$ are terms, and for any $X \in \mathcal{V}_\omega$, $X(t_1, \ldots, t_{\rho(X)})$ is also an atomic formula where $t_1, \ldots, t_{\rho(X)}$ are terms. Then, the second-order formulas are constructed from the atomic formulas together with the propositional connectives and quantifications over first-order and relational variables.

The set of all second-order formulas over a relational signature $\mathcal{R}$ with free variables in $\mathcal{X} \subseteq \mathcal{V}_\omega$ is denoted as $\text{SO}(\mathcal{R}, \mathcal{X})$. In the following we will represent a formula $\varphi \in \text{SO}(\mathcal{R}, \{X_1, \ldots, X_m, x_1, \ldots, x_n\})$ as $\varphi(X_1, \ldots, X_m, x_1, \ldots, x_n)$ to specify its free variables. For a structure $S \in \text{STR}(\mathcal{R})$ given by $S = \langle D_S, (R_S)_{R\in\mathcal{R}_+}, (c_S)_{c\in\mathcal{R}_0}\rangle$, for a formula $\varphi \in \text{SO}(\mathcal{R}, \{X_1, \ldots, X_m, x_1, \ldots, x_n\})$, $E_1 \subseteq D_S^{\rho(X_1)}, \ldots, E_m \subseteq D_S^{\rho(X_m)}$ and $d_1, \ldots, d_n \in D_S$, $S \models \varphi(E_1, \ldots, E_m, d_1, \ldots, d_n)$ denotes that φ is true in S for the values $E_1, \ldots, E_m$ of $X_1, \ldots, X_m$ and $d_1, \ldots, d_n$ of $x_1, \ldots, x_n$.

Let V_1 be a set of the countable set V_0 of first-order variables and the countably many relation variables of arity one from V_ω. A *monadic second-order formula* is a second-order formula written with variables from V_1. The set of all monadic second-order formulas with free variables in $X \subseteq V_1$, over a relational signature R, is denoted by $MSOL(R, X)$.

Note that any relation variable can be interpreted as a set. Thus, any relation variable will be called a *set variable*.

5.4.4 MONADIC SECOND-ORDER LOGIC DECISION AND OPTIMIZATION PROBLEMS

The following definitions are from [28].

Definition 5.14   A decision problem is an $MSOL(R)$ decision problem over $STR(R)$, if it can be expressed in the following form: Given an R-structure $S \in STR(R)$ and an $MSOL(R)$ closed formula φ, does $S \models \varphi$ hold?

Example 1: Let $\mathbf{G} = (V, E)$ be graph and let $R_s := \{edg\}$ be a relational signature with $\rho(edg) = 2$. An R_s-structure for $\mathbf{G}$ is defined by $\lfloor \mathbf{G} \rfloor := \langle V_\mathbf{G}, edg_\mathbf{G} \rangle$ where $edg_\mathbf{G} \subseteq V_\mathbf{G}^{[2]}$ such that $\forall u, v \in V : [\{u, v\} \in edg_\mathbf{G} \Leftrightarrow \{u, v\} \in E]$ and $V_\mathbf{G}$ is the vertex set of $\mathbf{G}$.

The 3-colorability problem is an $MSOL(R_s)$ decision problem since it can be stated as follows: Let $\mathbf{G}$ be a graph and let $\lfloor \mathbf{G} \rfloor$ be a structure for $\mathbf{G}$, does $\lfloor \mathbf{G} \rfloor \models \gamma_3$? where γ_3 is the closed $MSOL(R_s)$ formula defined as

$$
\begin{aligned}
\gamma_3 \;=\; \exists X_1, X_2, X_3 \Big(& Part(X_1, X_2, X_3) \wedge \\
& \forall u, v \big(edg_\mathbf{G}(u, v) \wedge u \neq v \Rightarrow \neg(X_1(u) \wedge X_1(v)) \wedge \\
& \neg(X_2(u) \wedge X_2(v)) \wedge \neg(X_3(u) \wedge X_3(v)) \big) \Big)
\end{aligned} \tag{5.31}
$$

and $Part(X_1, X_2, X_3)$ is defined as

$$
\begin{aligned}
Part(X_1, X_2, X_3) \;=\; & \forall v \Big(\big(X_1(v) \vee X_2(v) \vee X_3(v) \big) \wedge \big(\neg(X_1(v) \wedge X_2(v)) \wedge \\
& \neg(X_2(v) \wedge X_3(v)) \wedge \neg(X_1(v) \wedge X_3(v)) \big) \Big).
\end{aligned}
$$

Thus, $\lfloor \mathbf{G} \rfloor \models \gamma_3$ if and only if $\mathbf{G}$ is 3-colorable.

Definition 5.15   An optimization problem P is said to be a $LinEMSOL(R)$ optimization problem over $STR(R)$, if it can be expressed in the following form: Given $S \in STR(R)$ and m evaluation functions $f_1, \ldots, f_m$ associating values to the elements of S, find relations $E_1, \ldots, E_l \subseteq Dom(S)$ for the free variables of the formula $\varphi(X_1, \ldots, X_l) \in MSOL(R, \{X_1, \ldots, X_l\})$ such that:

$$
\sum_{\substack{1 \le i \le l \\ 1 \le j \le m}} a_{ij} E[X_i]_j = opt \Big\{ \sum_{\substack{1 \le i \le l \\ 1 \le j \le m}} a_{ij} E'[X_i]_j : S \models \varphi(E_1', \ldots, E_l') \Big\}, \tag{5.32}
$$

where $E[X_i]_j := \sum_{b \in E_i} f_j(b)$, *opt* is either min or max and $\{a_{ij} : 1 \leq i \leq l, 1 \leq j \leq m\}$ is a set of ml integers.

Example 2: Let $\mathbf{G} = (V, E, \delta)$ be a weighted graph where $\delta : V \to \mathbb{Z}$ is a map. For a subset $A \subseteq V$, the weight of A is defined as $w(A) := \sum_{a \in A} \delta(a)$. The *maximum weighted clique problem* (MWC) consists in finding a clique in $\mathbf{G}$ with maximum weight.

The MWC problem can be expressed as follows. Find an instance relation $V_1 \subseteq V_{\mathbf{G}}$ to the free variable X_1 in θ such that:

$$\sum_{a \in V_1} f_1(a) = \max\left\{ \sum_{a \in V_1'} f_1(a) : \lfloor \mathbf{G} \rfloor \models \theta(V_1') \right\} \tag{5.33}$$

where

$$\theta(X_1) = \forall u, v((X_1(u) \wedge X_1(v) \wedge u \neq v) \Rightarrow edg_{\mathbf{G}}(u, v)) \tag{5.34}$$

and $f_1 : V \to \mathbb{Z}$ is a function with $a \mapsto \delta(a)$.

Remark 5.16   Every MSOL($\mathcal{R}$) decision problem can be expressed as a LinEMSOL($\mathcal{R}$) optimization problem.

Note that, in order to optimize the objective function (5.33), every instance relation $V_1' \subseteq V_{\mathbf{G}}$ must satisfies that $\lfloor \mathbf{G} \rfloor \models \theta(V_1')$. It is possible to define another objective function as in (5.33) without restrictions if a penalty function is added. Consider the following definition.

Let P be a LinEMSOL optimization problem as defined in (5.32), then there is an MSOL expression $\varphi(X_1, \ldots, X_l)$ over a structure S, and evaluation functions $f_1, \ldots, f_m$, and a set of integers $\{a_{ij} : 1 \leq i \leq l, 1 \leq j \leq m\}$. For any instance relations $E_1, \ldots, E_l \subseteq \text{Dom}(S)$, let us define

$$\Gamma_P(E_1, \ldots, E_l) = \sum_{\substack{1 \leq i \leq l \\ 1 \leq j \leq m}} a_{ij} E[X_i]_j + g_\varphi(E_1, \ldots, E_l) \tag{5.35}$$

where g_φ is a penalty function such that $g_\varphi(E_1, \ldots, E_l)$ is equal to some constant c if and only if $S \models \varphi(E_1, \ldots, E_l)$, and $g_\varphi(E_1, \ldots, E_l) \gg c$ otherwise.

Remark 5.17   Minimizing (5.32) is equivalent to minimizing (5.35). A similar result can be stated for maximization.

5.4.5   MSOL OPTIMIZATION PROBLEMS AND PSEUDO-BOOLEAN MAPS

A *pseudo-Boolean map* $f : Q^n \to \mathbb{R}^+$ on n Boolean variables is a non-negative real-valued map on the hypercube Q^n. From Theorem 5.1, f can be represented as:

$$f(X) = \sum_{S \in \mathcal{P}(\llbracket 0, n-1 \rrbracket)} c(S) \prod_{j \in S} x_j$$

for some map $c : \mathcal{P}(\llbracket 0, n - 1 \rrbracket) \to \mathbb{R}$.

The set of Boolean variables will be denoted by $X = \{x_i : 0 \le i \le n - 1\}$ and the set of literals will be denoted by $L = \{x_i, \bar{x}_i : 0 \le i \le n - 1\}$ where $\bar{x}_i := 1 - x_i$.

A *disjunctive form* (DF) is an expression of the form

$$\phi = \bigvee_{k=1}^{m} \left(\bigwedge_{i \in A_k} x_i \wedge \bigwedge_{j \in B_k} \bar{x}_j \right) \tag{5.36}$$

where $A_k \cap B_k = \emptyset$ for $k = 1, \ldots, m$.

A DF ϕ is said to be *orthogonal* if $(A_k \cap B_l) \cup (A_l \cap B_k) \ne \emptyset$ for all $k, l \in \{1, \ldots, m\}$ with $k \ne l$.

It is said that a DF ϕ *represents* a Boolean function g if the true valued points of g coincide with the true valued points of ϕ.

Theorem 5.18 Every Boolean function $g : Q^n \to Q$ can be represented through an orthogonal DF ϕ_g.

See a proof in [31].

Theorem 5.19 Every Boolean function $g : Q^n \to Q$ represented by an orthogonal DF, has an associated multilinear polynomial given as

$$g(X) = \sum_{k=1}^{m} \left(\prod_{i \in A_k} x_i \prod_{j \in B_k} (1 - x_j) \right).$$

See a proof in [31].

From Theorem 5.19 it follows that every FO sentence has an associated multilinear polynomial. Let us consider the following algorithm to obtain a multilinear polynomial of a given FO sentence.

FO sentence into multilinear polynomial (1)

Input: A FO(Σ) sentence φ where $\Sigma = (\Phi, \Pi)$.

Output: A multilinear polynomial p_φ over a set of Boolean variables X.

1. Transform φ into a DF (see [86] for a standard procedure).

2. Apply the equivalences $x \wedge y = xy$, $x \vee y = x + y - xy$ and $\bar{x} = 1 - x$ on φ to obtain an arithmetic expression p_φ.

3. For every atomic formula s in p_φ introduce a Boolean variable X_s to obtain a multilinear polynomial over $X = \{X_s | s$ is an atomic formula in $\varphi\}$.

Step 1 in the algorithm 1 drop all existential quantifier in the sentence φ by Skolemization.

Example 3: Let $\mathbf{G} = (V, E)$ be a undirected graph, and let

$$\varphi = \forall x (\neg edg(x, x)) \wedge \neg \exists w, x, y, z (edg(w, x) \wedge edg(x, y) \wedge edg(y, z) \\ \wedge \neg edg(w, y) \wedge \neg edg(w, z) \wedge \neg edg(x, z)).$$

Then, it is satisfied that $\lfloor \mathbf{G} \rfloor \models \varphi$ if and only if $\mathbf{G}$ has no loops and no induced subgraph isomorphic to P_4 (P_4 is the graph $\bullet - \bullet - \bullet - \bullet$).

By transforming φ into a DNF we obtain

$$\varphi = (\neg edg(u, u) \wedge \neg edg(w, x)) \vee (\neg edg(u, u) \wedge \neg edg(x, y)) \vee \\ (\neg edg(u, u) \wedge \neg edg(y, z)) \vee (\neg edg(u, u) \wedge edg(w, y)) \vee \\ (\neg edg(u, u) \wedge edg(w, z)) \vee (\neg edg(u, u) \wedge edg(x, z)),$$

and after applying steps 2 and 3 of the algorithm (1)

$$p_\varphi = \left(\sum_{u \in V} (1 - X_{uu}) \right) \sum_{w, x, y, z \in V} (1 - X_{wx} X_{xy} X_{yz} + X_{wx} X_{xy} X_{yz} X_{wy} + \\ X_{wx} X_{xy} X_{yz} X_{wz} - X_{wx} X_{xy} X_{yz} X_{xz} - X_{wx} X_{xy} X_{yz} X_{wy} X_{wz} - \\ X_{wx} X_{xy} X_{yz} X_{wy} X_{xz} - X_{wx} X_{xy} X_{yz} X_{wz} X_{xz} + X_{wx} X_{xy} X_{yz} X_{wy} X_{wz} X_{xz}),$$

where $\forall u, v \in V : X_{uv} := edg(u, v)$.

Finally, given p_φ as input to the procedure Reduce in Section 5.1, it produces a quadratic pseudo-Boolean map.

Remark 5.20   Given a undirected graph $\mathbf{G} = (V, E)$, $p_\varphi(X) = (card\ V)^5$ if and only if, $\mathbf{G}$ has no loops and no induced subgraph isomorphic to P_4. $p_\varphi(X) < (card\ V)^5$ otherwise.

In the second-order logic it is allowed to quantify over relations of any arity, then in order to obtain a polynomial expression of a given SO sentence ψ, we will write ψ in existential second-order logic form.

An sentence of *existential second-order logic* (ESOL) ψ over a signature $\Sigma = (\Phi, \Pi)$ is of the form

$$\psi = \exists R_1 \cdots \exists R_r \varphi,$$

where $R_1, \dots, R_r$ are relational symbols of respective arities $\rho(R_1), \dots, \rho(R_r)$ and φ is a first-order sentence over the signature $\Sigma' = (\Phi, \Pi \cup \{R_1, \dots, R_r\})$. A structure $S \in STR(\Sigma)$ satisfies an ESOL sentence $\exists R_1 \cdots \exists R_r \varphi$, if there are relations $E_1 \subseteq D_S^{\rho(R_1)}, \dots, E_r \subseteq D_S^{\rho(R_r)}$ such that S, augmented with $\{E_1, \dots, E_r\}$ to comprise a structure for Σ', satisfies φ.

An ESOL sentence $\exists R_1 \cdots \exists R_r \varphi$ is an *existential monadic second-order logic* (EMSOL) sentence if the relations $R_1, \dots, R_r$ are of arity one.

Theorem 5.21 Fagin, 74 Every decision problem on finite graphs is in NP if and only if it is expressible in existential second-order logic.

Examples of graph problems that can be written as an EMSOL expression are 3-colorability, Maximum Clique problem, Hamiltonian circuit problem, and TSP problem.

Let us consider the following algorithm to obtain a multilinear polynomial of a given EMSOL expression.

EMSOL sentence into multilinear polynomial (2)

Input: A $SO(\Sigma)$ sentence ψ where $\Sigma = (\Phi, \Pi)$.

Output: A multilinear polynomial p_ψ over a set of Boolean variables X.

1. Transform ψ into an EMSOL sentence $\exists R_1, \ldots, R_r.\varphi$.

2. Transform φ into a DF (see [86] for a standard procedure).

3. Apply the equivalences $x \wedge y = xy, x \vee y = x + y - xy$ and $\bar{x} = 1 - x$, on φ to obtain an arithmetic expression p_φ.

4. For every atomic formula s in p_φ introduce a Boolean variable X_s to obtain a multilinear polynomial p_ψ over $X = \{X_s | s$ is an atomic formula in $\varphi\}$.

Let us consider the following example:

Example 4: Let $\mathbf{G} = (V, E)$ be a simple undirected graph, and let

$$
\begin{aligned}
\gamma_3 \;=\; \exists X, Y, Z \Big(& Part(X, Y, Z) \wedge \\
& \forall u, v \big(edg_{\mathbf{G}}(u, v) \Rightarrow \neg(X(u) \wedge X(v)) \wedge \\
& \neg(Y(u) \wedge Y(v)) \wedge \neg(Z(u) \wedge Z(v))\big)\Big)
\end{aligned}
$$

and $Part(X, Y, Z)$ is defined by

$$
\begin{aligned}
Part(X, Y, Z) \;=\; \forall v\Big(& \big(X(v) \vee Y(v) \vee Z(v)\big) \wedge \big(\neg(X(v) \wedge Y(v)) \wedge \\
& \neg(Y(v) \wedge Z(v)) \wedge \neg(X(v) \wedge Z(v))\big)\Big).
\end{aligned}
$$

Then, it is satisfied that $\lfloor \mathbf{G} \rfloor \models \gamma_3$ if and only if, $\mathbf{G}$ is 3-colorable.

It can be seen that γ_3 is already in EMSOL form, then we can apply the algorithm (2). We claim that there are polynomials p_{φ_1} and p_{φ_2} such that $p_\psi = p_{\varphi_1} \cdot p_{\varphi_2}$, where

$$
\begin{aligned}
p_{\varphi_1} \;=\; \sum_{u \in V} & (X_u + Y_u + Z_u - X_u Y_u - X_u Z_u - Y_u Z_u + X_u Y_u Z_u) \cdot \\
& (1 - X_u Y_u)(1 - Y_u Z_u)(1 - X_u Z_u)
\end{aligned} \tag{5.37}
$$

and

$$P_{\varphi_2} = \sum_{u,v \in V} \left((1 - X_{uv}) + (1 - X_u X_v)(1 - Y_u Y_v)(1 - Z_u Z_v) X_{uv}\right). \tag{5.38}$$

For any non-empty sets $X, Y, Z \subseteq V$, $p_\psi(X, Y, Z)$ is a multilinear polynomial over the set of Boolean variables $X = \{X_{uv}|u, v \in V\} \cup \{X_u|u \in X\} \cup \{Y_u|u \in Y\} \cup \{Z_u|u \in Z\}$.

Remark 5.22   Given a simple undirected graph $\mathbf{G} = (V, E)$ and any subsets $X, Y, Z \subseteq V$ then $\mathbf{G}$ is 3-colorable if and only if, $p_\psi(X, Y, Z) = (card\ V)^2$. $p_\psi(X, Y, Z) < (card\ V)^2$ otherwise.

In Example 4 the polynomial expressions given in (5.37) and (5.38) depend on the given partition (X, Y, Z). Then, it is not possible to state the optimization problem using the algorithm 2 for the 3-coloring problem. A possible alternative to construct a polynomial expression for the 3-coloring problem is by considering an objective function over all possible partitions such that a partition (X, Y, Z) is a 3-coloring if and only if, the objective function is minimized.

Proposition 5.23   Any existential monadic second order logic sentence is suitable to be expressed through a polynomial map defined on the hypercube.

Proposition 5.23 can be proved by considering the algorithm 2, but restricted to subset graph problems, i.e., the Maximum Clique Problem.

Remark 5.24   The polynomial expression obtained by algorithm 2 can be considered as a penalty function in (5.35).

The polynomial expression returned by algorithm 2 can be reduced to a quadratic form and subsequently to be optimized using an AQO algorithm. Also, from Remark 5.24 the objective function defined in (5.35) can also be used in AQO. These polynomial expressions provide us a general scheme to deal with optimization problems that are expressible in MSOL.

CHAPTER 6

A General Strategy to Solve NP-Hard Problems

In computational complexity, it is well known that NP-hard problems are the most difficult problems to solve and in many cases only an approximation to the optimal solution is given (see [9]). On the other hand, logical characterization of NP-problems has provided a classification of NP optimization problems in terms of first- and second-order logic expressions [5, 50, 61, 100]. *Tree-decomposition* and *treewidth* of graphs are important concepts introduced in a series of publications on graph minors [79–83]. In references [27–29] show that on graphs of bounded treewidth, every decision or optimization problem expressible in Monadic Second Order Logic (MSOL) has a linear time solution. Many references [14–17, 55, 60, 77, 93] show that there is a *dynamic programming approach* on tree-decomposition on graphs of bounded treewidth to solve optimization problems in linear time. Another important concept related to the treewidth is the *Clique-width* [28, 30, 43] which has been considered to show linear time algorithms on graphs of bounded clique-width. Recently, it has been considered the *Tree-Depth* as a parameter to build efficient algorithms [69].

This chapter is divided into two parts. The first part is devoted to a study on tree-decompositions; here, we analyze the iterative construction and updating of tree-decompositions, done by adding one edge at a time [33]. In the second part we consider the dynamic programming approach for solving optimization problems. We propose a solution to the *classical Ising spin glass model* based on the Dynamic Programming approach. We also propose a composition strategy of local Hamiltonians for AQC on tree-decompositions of graphs.

6.1 BACKGROUND

6.1.1 BASIC NOTIONS

Let $\mathbf{G} = (V, E)$ be a graph with $V(\mathbf{G}) = V$ as set of vertices and $E(\mathbf{G}) = E$ as set of edges, and let $n = |V|$ be the number of vertices, or *graph order*. For any subset $S \subseteq V$ of vertices, the subgraph of $\mathbf{G}$ *induced* over S, denoted by $\mathbf{G}[S]$, is the graph $\mathbf{S} = (S, E_S)$ where $E_S = \{\{x, y\} \in E \,|\, x, y \in S\}$. A *clique* in $\mathbf{G}$ is a complete subgraph of $\mathbf{G}$. The *clique number* $\omega(\mathbf{G})$ is the size of the largest clique in $\mathbf{G}$.

If $i < j$, then $[\![i, j]\!]$ will denote the set of integers $\{i, i + 1, \ldots, j - 1, j\}$, and from now on we will identify V with $[\![0, n - 1]\!]$.

A *Hamiltonian path* in $\mathbf{G}$ may be realized as a subgraph $\mathbf{P}_\pi = (V, E_\pi)$ of $\mathbf{G}$, where π is a permutation of V and $E_\pi = \{\{\pi(i), \pi(i+1)\}|0 \leq i < n-1\}$. The vertices $\pi(0)$ and $\pi(n-1)$ are *linked* by $\mathbf{P}_\pi$ and are called the *ending points* of the path.

A *Hamiltonian cycle* has the form $\mathbf{C}_\pi := \mathbf{P}_\pi + \{\pi(0), \pi(n-1)\}$ where $\mathbf{P}_\pi$ is a Hamiltonian path. In other words, a Hamiltonian cycle is a Hamiltonian path whose ending points form an edge.

A *cycle* is a Hamiltonian cycle in a subgraph of $\mathbf{G}$. The *length* of a cycle is the number of its edges. A *chord* is an incident edge to two vertices that are not adjacent within the cycle. The graph $\mathbf{G}$ is *triangulated* (or *chordal*) if every cycle of length at least 4 has a chord. A *triangulation* of $\mathbf{G}$ is a graph $\mathbf{H}$ with the same set of vertices such that $\mathbf{G}$ is a subgraph of $\mathbf{H}$ and $\mathbf{H}$ is triangulated, and it is a *minimal triangulation* of $\mathbf{G}$ if there is no triangulation of $\mathbf{G}$ that is a proper subgraph of $\mathbf{H}$.

Definition 6.1 The notion of k-*tree* is defined recursively as follows.

1. A clique with $k+1$ vertices is a k-tree.

2. Given a k-tree $\mathbf{T}_n$ with n vertices, it is expanded to a k-tree with $n+1$ vertices as follows: add a new vertex x_{n+1}, choose a k-clique of $\mathbf{T}_n$, and connect x_{n+1} with each vertex in the chosen k-clique.

A *partial k-tree* is a subgraph of a k-tree with the same set of vertices. The *treewidth* of a graph $\mathbf{G}$ is the minimum value k for which $\mathbf{G}$ is a partial k-tree. Any k-tree has treewidth k.

Problem 6.2 Treewidth Problem Given a graph $\mathbf{G}$ and an integer $k \geq 1$, decide whether the treewidth of $\mathbf{G}$ is at most k.

As was shown in [6], the treewidth problem is NP-complete. However, the Treewidth Problem restricted to graphs with treewidth bounded by a parameter $k_b \in \mathbb{Z}^+$, is decidable in linear time [60].

A simple characterization of k-trees was shown in [84].

Lemma 6.3 [84] *A graph $\mathbf{G}$ with n vertices is a k-tree if and only if $\mathbf{G}$ is triangulated, $\omega(\mathbf{G}) = k+1$, and $|E(\mathbf{G})| \geq nk - \frac{1}{2}k(k+1)$.*

A characterization of triangulated graphs was shown in [40]. Let us recall it. Let $\mathbf{G} = (V, E)$ be a graph, and let $x \in V$. The vertex x is *simplicial* if the subgraph induced by $\mathbf{G}$ over the neighborhood $N(x) := \{y \in V|\{x, y\} \in E\}$ is a clique. Let σ be a permutation of V. For an index $i \in [\![0, n-1]\!]$, let $\mathbf{G}[\sigma(i, n)]$ denote the subgraph $\mathbf{G}[S_i]$ induced by $\mathbf{G}$ over $S_i = \{\sigma(i), \ldots, \sigma(n-1)\}$. It is said that σ is a *perfect elimination scheme* (PES) in $\mathbf{G}$ if for each $i \in [\![0, n-1]\!]$, the vertex $\sigma(i)$ is simplicial in $\mathbf{G}[\sigma(i, n)]$.

The triangulated graphs are determined as follows.

Lemma 6.4 [40] *A graph **G** is triangulated if and only if there exists a PES for **G**. Furthermore, if a graph is triangulated, any simplicial vertex can start a PES for the graph.*

Now let us recall the *Minimum Triangulation Problem*. Let σ be a permutation of the vertex set V. The *fill-in* produced by σ, denoted $Fill(\sigma)$, is a set of new edges that should be added to the graph **G** in such a way that for each $i \in [\![0, n-1]\!]$, the vertex $\sigma(i)$ becomes simplicial in $\mathbf{G}[\sigma(i, n)]$. Consequently, if σ is a PES then $Fill(\sigma) = \emptyset$.

Problem 6.5 Minimum Fill-in Given a graph $\mathbf{G} = (V, E)$, find a permutation σ of V such that $Fill(\sigma)$ is minimum.

Equivalently, the Minimum Fill-in Problem can be stated as finding the minimum set of edges whose addition to the graph **G** is a chordal graph. The Minimum Fill-in Problem is indeed NP-complete [97], however the decision of whether a given graph **G** is triangulated, can be done in linear time with respect to the number of vertices [47, 59, 68, 85].

The algorithm 1 Fill-in($\mathbf{G}, \pi$) below receives a graph **G** and a permutation π of the set of vertices, and constructs a triangulation **H** of **G** such that π is a PES in **H**, by adding a minimum number of edges to **G**. Clearly, the time complexity of this algorithm is of the order $O(nm)$, where $m = |E|$ is the number of edges of the input graph.

Algorithm 1 Fill-in($\mathbf{G}, \pi$)

Require: A graph $\mathbf{G} = (V, E)$ with $n = |V|$ and a permutation $\pi : [\![0, n-1]\!] \to V$.
Ensure: A triangulation **H** of **G** such that π is a PES for **H**.
  $\mathbf{H} := \mathbf{G}$;
  for all $i = 0, \ldots, n-1$ **do**
    Let $v = \pi(i)$ be the i-th vertex according to π;
    for all pair $w, u \in N(v)$ such that $\pi^{-1}(w) > i$, $\pi^{-1}(u) > i$ **do**
      if w and u not adjacent in **H then**
        Add $\{w, u\}$ to **H**
      end if
    end for
  end for
  Return **H**.

6.1.2 TREE DECOMPOSITIONS

Definition 6.6 A tree decomposition of a graph $\mathbf{G} = (V, E)$ is a pair $(\mathbf{T}, \mathcal{X})$ where $\mathbf{T} = (T, F)$ is a tree, and $\mathcal{X} = (X_t)_{t \in T}$ is a family of subsets of V such that the following conditions are satisfied:

1. $\bigcup_{t \in T} X_t = V$,

2. $\forall \{u, v\} \in E \; \exists t \in T \colon u, v \in X_t$, and

3. $\forall x \in V$ the subgraph induced by $\mathbf{T}$ over $\{t \in T \,|\, x \in X_t\}$ is a subtree of $\mathbf{T}$.

Alternatively, condition 3 can be formulated as follows.

3'. *For all $t_1, t_2, t_3 \in T$, if t_2 is on the path connecting t_1 with t_3 in $\mathbf{T}$ then $X_{t_1} \cap X_{t_3} \subset X_{t_2}$.*

For each tree vertex $t \in T$, the subset $X_t \subset V$ of graph vertices is called its *bag*. The *width* of a tree decomposition $(\mathbf{T}, \mathcal{X})$ is $\max_{t \in T} |X_t| - 1$. The *treewidth* of a graph $\mathbf{G}$ is the minimum width over all possible tree decompositions of $\mathbf{G}$, and it is written as $\mathrm{tw}(\mathbf{G})$.

Definition 6.7 A *branch decomposition* of a graph $\mathbf{G} = (V, E)$ is a tree decomposition $(\mathbf{T}, \mathcal{X})$ such that $\mathbf{T}$ is just a branch, namely a path.

The *branchwidth* of the graph $\mathbf{G}$ is the minimum width over all possible branch decompositions of $\mathbf{G}$.

Algorithm 2 below produces a tree decomposition of $\mathbf{G}$, with the same set of vertices, assuming that a PES π in $\mathbf{G}$ is given. For each vertex at $\mathbf{G}$, it is required the computation of the subgraph $\mathbf{G}[\pi(k, n)]$, and then an exploration on the edges is necessary in order to find the index j in the main cycle of the algorithm, thus, the time complexity of the algorithm is of the order $O(n^2 m)$.

Algorithm 2 GeneralTreeDecomposition$(\mathbf{G}, \pi)$

Require: A graph $\mathbf{G} = (V, E)$, $n = |V|$, and a PES $\pi : [\![0, n - 1]\!] \to V$ of $\mathbf{G}$.
Ensure: A tree decomposition $(\mathbf{T} = (V, F), \mathcal{X})$ of $\mathbf{G}$.
  Let $((T, F), \mathcal{X}) = ((V, \emptyset), \emptyset)$ be the initial empty tree decomposition;
  for all $k = n - 1, \ldots, 0$ **do**
    if $k == n - 1$ **then**
      $X_{\pi(k)} = \{\pi(k)\}$;
    else
      Let $\mathbf{G}' = (V', E') := \mathbf{G}[\pi(k, n)]$;
      Let $\pi(j)$ be the lowest numbered neighbor of $\pi(k)$ in $\mathbf{G}'$, i.e.,
      $j := \min\{i \in [\![0, n - 1]\!] \,|\, \{\pi(k), \pi(i)\} \in E'\}$;
      Let $X_{\pi(k)} := N(\pi(k), \mathbf{G}') \cup \{\pi(k)\}$: the neighborhood of $\pi(k)$ in $\mathbf{G}'$;
      Let $F := F \cup \{\pi(k), \pi(j)\}$;
    end if
  end for
  Return $(\mathbf{T} := (V, F), \mathcal{X})$.

In order to produce a tree decomposition of a graph in a general setting, with Algorithm 1, and any permutation π, a triangulation $\mathbf{H}$ is produced with π as PES and then Algorithm 2 produces the tree decomposition. Since a triangulation of a graph with n vertices will have $O(n)$ edges, the composition of Algorithm 1 with Algorithm 2 has time complexity $O(n^2 m)$.

6.2   PROCEDURAL MODIFICATION OF TREE DECOMPOSITIONS

Let us consider a general algorithm to construct a tree decomposition using elimination schemes. Given a graph $\mathbf{G} = (V, E)$, a tree decomposition $(\mathbf{T}, \mathcal{X})$ of $\mathbf{G}$ is obtained as follows:

$$(\mathbf{G}, \pi) \xmapsto{\ \Psi\ } \mathbf{H}_\pi \xmapsto{\ \Phi\ } (\mathbf{T}, \mathcal{X}), \tag{6.1}$$

where Ψ is a procedure to triangulate $\mathbf{G}$ in such a way that π is a PES in $\mathbf{H}_\pi$ (for instance, Algorithm 1), and Φ is the transformation calculated by Algorithm 2.

Any tree decomposition obtained using the general procedure (6.1), depends in the vertex ordering determined by π. In general, for a fixed triangulation, different PES's will produce different tree decompositions.

6.2.1   MODIFICATION BY THE ADDITION OF AN EDGE

Now, suppose that we have an already constructed tree decomposition $(\mathbf{T}, \mathcal{X})$ of a graph $\mathbf{G}$. Let us modify the graph by the addition of an edge. Let $\mathbf{G} + e$ be the new graph, and let us pose as a task to build a corresponding tree decomposition $(\mathbf{T}', \mathcal{X}')$ of $\mathbf{G} + e$, see the following diagram:

$$\begin{array}{ccc} \mathbf{G} & \xmapsto{\ A\ } & \mathbf{G} + e \\ {\scriptstyle \Phi \circ \Psi}\downarrow & & {\scriptstyle \Phi \circ \Psi} \\ (\mathbf{T}, \mathcal{X}) \rightsquigarrow_B (\mathbf{T}', \mathcal{X}') & & (\mathbf{T},'' \mathcal{X}'') \end{array} \tag{6.2}$$

where $(\Phi \circ \Psi)$ is the general algorithm sketched at diagram (6.1), A is the addition of an edge to a graph, and B is the sought corresponding tree decomposition transformation. Naturally, the procedure $(\Phi \circ \Psi)$ can be applied to the modified graph $\mathbf{G} + e$, producing thus a tree decomposition $(\mathbf{T},'' \mathcal{X}'')$. Both trees $(\mathbf{T}', \mathcal{X}')$ and $(\mathbf{T},'' \mathcal{X}'')$ are tree decompositions of the graph $\mathbf{G} + e$.

It is important to note that, when an edge e is added to $\mathbf{G}$, it is not generally true that the current triangulation $\mathbf{H}_\pi = \Psi(\mathbf{G})$ remains a triangulation of the modified graph $\mathbf{G} + e$, thus it would be necessary to construct a new triangulation for $\mathbf{G} + e$. However, Lemma 6.4 asserts that there exists a PES which can be used to test whether $\mathbf{H}_\pi + e$ is triangulated or not. In the affirmative case, no new computation of the triangulation is required.

The next section gives some results about this approach.

6.2.2   ITERATIVE MODIFICATION

Let $\mathbf{G} = (V, E)$ be a graph with $n = |V|$ vertices and let $\mathbf{H}$ be a triangulation of the graph $\mathbf{G}$. By the lemma 6.4, there exists a PES π in $\mathbf{H}$. For each index $i \in [\![0, n-1]\!]$, let $N_{\pi(i)}$ be the neighborhood of $\pi(i)$ in $\mathbf{H}[\pi(i, n)]$. Then $\mathbf{H}[\pi(i, n)]$ induces a clique over $N_{\pi(i)}$.

Remark 6.8   Let $e = \{u, v\}$ be an edge not in $\mathbf{H}$ and let $j_1 := \min\{\pi^{-1}(u), \pi^{-1}(v)\}$ and $j_2 := \max\{\pi^{-1}(u), \pi^{-1}(v)\}$. Then the following conditions are equivalent.

1. $\mathbf{H}' := \mathbf{H} + e$ is triangulated and it has π as a PES.

2. $\mathbf{H}'[\pi(j_1, n)]$ induces a clique over $N_{\pi(j_1)} \cup \{\pi(j_2)\}$.

Let us say that any edge $e \in E$ *maintains the triangulation* if condition 1 in Remark 6.8 holds. In a procedural way, the checking of whether an edge maintains the triangulation can be done, through condition 2, in time complexity $O(\text{degree}_{\mathbf{H}}(j_1)^2) \le O(|E|^2)$.

Hence, if an edge e is added to a triangulated graph $\mathbf{H}_\pi$, and π remains as a PES for $\mathbf{H}_\pi + e$, then the neighborhood $N_{\pi(j)}$ of just one vertex $\pi(j)$ increases by one element, while the other neighborhoods do not change.

Claim 6.9   Let $\mathbf{H}$ be a triangulation of a graph $\mathbf{G}$, different than the whole clique K_V, and let π be a PES in $\mathbf{H}$. Then, there exists an edge e, not in $E(\mathbf{G})$, that maintains the triangulation.

The selection of such an edge e can be applied iteratively until the arrival to the whole clique K_V.

Proof. Let $e \in E$ be an edge not in the triangulated graph $\mathbf{H}_\pi$, and let j_1, j_2 be two indexes defined as in remark 6.8. Let j be an index such that $j_1 < j$ and $\pi(j) \in N_{\pi(j_1)}$, and let $N' = \{\pi(j_2)\} \cup (N_{\pi(j_1)} \backslash \{\pi(j)\})$. Then $\mathbf{H} + e$ remains triangulated while $N' \subset N_{\pi(j)}$, just because $\mathbf{H}[\pi(j, n)]$ induces a clique over N' and $\pi(j)$ is connected to N'. $\square$

The claim 6.9 is the basis of an iterative procedure: Choose a PES π for $\mathbf{H}$, then pick an edge $e \in E - E(\mathbf{H})$, and check whether π is a PES for $\mathbf{H}' := \mathbf{H} + e$, in which case update $\mathbf{H}'$ as the triangulated graph. Repeat the procedure.

Now, let $(\mathbf{T} = (V, F), \mathcal{X})$ be a tree decomposition of a graph $\mathbf{G} = (V, E)$ obtained using the general algorithm $\Phi \circ \Psi$, then there exists a triangulation $\mathbf{H}_\pi$ of $\mathbf{G}$, for a PES π in $\mathbf{H}$, such that Ψ produces $(\mathbf{T}, \mathcal{X})$ from $\mathbf{H}_\pi$.

Remark 6.10   Let $e = \{u, v\}$ be an edge such that $\mathbf{H}_\pi + e$ is triangulated. Let $(\mathbf{T}' = (V, F'), \mathcal{X}')$ be its tree decomposition as defined in algorithm 2. Using the notation in Remark 6.8, let $\pi(j)$ be the lowest numbered neighbor of $\pi(j_1)$ in $\mathbf{H}_\pi[\pi(j_1, n)]$. Then, initially, $(\mathbf{T}', \mathcal{X}') = (\mathbf{T}, \mathcal{X})$ and consecutively it is modified according to the following cases.

1. If $j < j_2$ then $\tau'(\pi(j_1)) := \tau'(\pi(j_1)) \cup \{\pi(j_2)\}$.

2. If $j > j_2$ then $\tau'(\pi(j_1)) := \tau'(\pi(j_1)) \cup \{\pi(j_2)\}$, $F' := F'\backslash\{\pi(j_1), \pi(j)\}$ and $F' := F' \cup \{\pi(j_1), \pi(j_2)\}$.

From Remark 6.10, it is easy to see that the tree decomposition $\mathbf{T}'$ does not change by adding new edges that maintain the triangulation and satisfies case 1, while the family of bags $\mathcal{X}'$

grows in one graph vertex at just one tree vertex when a new edge is added. Also, it is important to note that the width of the tree $\mathbf{T}'$ increases in one only when the cardinality of the bag at vertex $\pi(j_1)$ is the greatest in $\mathbf{T}$.

If an edge e does not maintain the triangulation, due to claim 6.9 there exists a sequence of edges $e_1, \ldots, e_k$ such that for the sequence of triangulated graphs $(\mathbf{H}_i)_{i=0}^{k}$, with $\mathbf{H}_0 = \mathbf{H}$ and $\mathbf{H}_i = \mathbf{H}_{i-1} + e_i$, the edge e_i maintains the triangulation $\mathbf{H}_{i-1}$ and the last edge e_k coincides with the original edge e. Let us say that the edge sequence $e_1, \ldots, e_k$ is a *climbing sequence* for e.

Then, the following problem can be posed.

Problem 6.11   For a given edge e that does not maintain the triangulation $\mathbf{H}$, find the minimum length among all possible climbing sequences for e.

Algorithm 3 below solves this problem by adding the necessary edges to the triangulation $\mathbf{H}$, in such a way that $\mathbf{H}[\pi(j_1, n)]$ induces a clique over $N_{\pi(j_1)} \cup \{\pi(j_2)\}$ and by repeating the same task for each element in $N_{\pi(j_1)}$.

We see that the array A acts as a queue in order to perform a breadth-first examination of potential edges in a climbing sequence. Hence, the time complexity of this algorithm is of the order $O(m)$, where m is the number of edges in the input triangulation.

6.2.3   BRANCH DECOMPOSITIONS

An application of the iterative modification of tree decompositions is the following. Let $\mathbf{G}$ be a graph, $\mathbf{H}_\pi = \Psi(\mathbf{G})$ be a triangulation of $\mathbf{G}$ and let $(\mathbf{T}, \mathcal{X}) = \Phi \circ \Psi(\mathbf{G})$ be its tree decomposition as defined in Algorithm 2 from $\mathbf{H}_\pi$. Then, it is possible to obtain a branch decomposition from $(\mathbf{T}, \mathcal{X})$ by adding new edges to $\mathbf{H}_\pi$ in order to maintain the triangulation with respect to π.

The idea behind this transformation from a tree to a branch decompositions is a consequence of the case 2 in Remark 6.10. Namely, when the condition 2 is fulfilled, the tree shrinks. Then, by adding the necessary edges to the triangulation $\mathbf{H}_\pi$ satisfying the condition 2, the tree becomes a branch.

Claim 6.12   Let $\mathbf{G} = (V, E)$ be a graph with $n = |V|$, $\mathbf{H}_\pi = \Psi(\mathbf{G})$ be a triangulation of $\mathbf{G}$ and let $(\mathbf{T}, \mathcal{X}) = \Phi \circ \Psi(\mathbf{G})$ be its tree decomposition as defined in Algorithm 2. Then, there exists a set of edges $\{e_1, \ldots, e_k\}$ such that when adding to $\mathbf{H}_\pi$ successively (satisfying Remark 6.10), $(\mathbf{T}, \mathcal{X})$ becomes a branch decomposition $(\mathbf{T}', \mathcal{X}')$ where $\mathbf{T}' = (V, F')$, $F' = \{\{\pi(i), \pi(i+1)\} | 0 \le i \le n-2\}$ and $\mathcal{X}'$ is described according to Remark 6.10.

Proof. By case 2 in Remark 6.10, when a new edge e is added to $\mathbf{H}_\pi$, the edge $\{\pi(j_1), \pi(j)\}$ is deleted from the tree and replaced by the edge $\{\pi(j_1), \pi(j_2)\}$ where the index $j_2 < j$. Hence, the lowest index j' such that $j' < j$ satisfies $j' = j_1 + 1$, and it corresponds indeed to the edge $\{\pi(j_1), \pi(j_1 + 1)\}$. □

Algorithm 3 MinimumClimbing

Require: An edge $e = \{u, v\}$, a triangulation **H** of a graph and a PES π for N.
Ensure: The length of the minimal climbing sequence for e.
 Let $A := \emptyset$; Let j_1, j_2 be defined as in remark 6.8;
 for all $w \in N_{\pi(j_1)}$ **do**
   if w is not adjacent to $\pi(j_2)$ **then**
     Add $\{w, \pi(j_2)\}$ to N; $A = A \cup \{w\}$;
   end if
 end for
 $B = A$;
 while A is not empty **do**
   Let $a \in A$ and $j = \pi^{-1}(a)$;
   for all $w' \in N_{\pi(j)}$ **do**
     if w' is not adjacent to $\pi(j_2)$ **then**
       Add $\{w', \pi(j_2)\}$ to **H**;
       $B = B \cup \{w'\}; A = A \cup \{w'\}$
     end if
   end for
   $A = A - \{a\}$;
 end while
 Return $|B|$.

Remark 6.13 Let $\mathbf{G} = (V, E)$ be a graph, $\mathbf{H}_\pi$ be a triangulation of $\mathbf{G}$ and let $(\mathbf{T}, \mathcal{X})$ be its tree decomposition as defined in Algorithm 2. By Claim 6.9 it is possible to arrive to the whole clique K_V, adding edges successively to $\mathbf{H}_\pi$. Then, $(\mathbf{T}, \mathcal{X})$ evolves into a branch decomposition of the complete graph K_V.

Remark 6.13 determines an upper bound of the number of required edges to transform a tree decomposition into a branch decomposition.

Then, the following problem can be posed.

Problem 6.14 For a given tree decomposition $(\mathbf{T}, \mathcal{X})$ of a graph $\mathbf{G}$, find the minimum number of edges satisfying the claim 6.12, in order to transform $(\mathbf{T}, \mathcal{X})$ into a branch decomposition.

Let us consider the following example: Let $\mathbf{G} = (V, E)$ be a graph where:

$$V = \{0, 1, 2, 3, 4, 5, 6\} \text{ and}$$
$$E = \{\{0, 1\}, \{0, 6\}, \{2, 1\}, \{2, 3\}, \{2, 6\}, \{3, 5\}, \{3, 4\}, \{4, 5\}, \{5, 6\}\}.$$

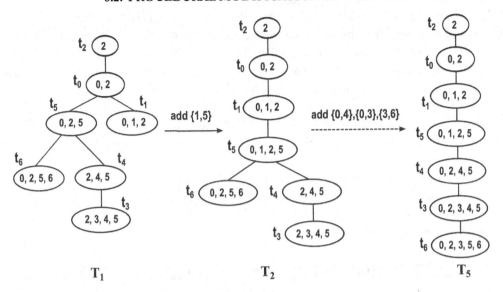

Figure 6.1: From left to right, a tree decomposition of a graph $G = (V, E)$. The modified tree decompositions T_2 until T_5 are shown when the edges $\{1, 5\}, \{0, 4\}, \{0, 3\}, \{3, 6\}$ are added to G. Each bag at vertex $v \in V$ is depicted as a set of vertices in V with label t_v.

Let $\pi := (6, 3, 4, 5, 1, 0, 2)$ be a permutation of V, Algorithm 1 with (G, π) as input returns a triangulation $H_1 = (V, E')$ of G where:

$$E' = E \cup \{\{0, 2\}, \{0, 5\}, \{2, 5\}, \{2, 4\}\}$$

and π is a PES for H_1. Using Algorithm 2 with (H_1, π) as input a tree decomposition T_1 is obtained. Note that π remains a PES for $H_2 := H_1 + \{1, 5\}$ and from Remark 6.10, T_1 becomes a tree decomposition T_2 that satisfy condition 2. In the same way, π remains a PES for H_3, H_4 where $H_3 := H_2 + \{0, 4\}$, $H_4 := H_3 + \{0, 3\}$, and its corresponding tree decompositions T_3 and T_4 that satisfy condition 1 in Remark 6.10.

Finally, π is a PES for $H_5 := H_4 + \{3, 6\}$ and T_4 becomes a branch decomposition T_5 that satisfies condition 2 in Remark 6.10.

Figure 6.1 shows the previous transformation.

Note also that $\{1, 5\}, \{0, 4\}, \{0, 3\}, \{3, 6\}$ are the minimum number of edges to transform T_1 into the branch decomposition T_5.

6.2.4   COMPARISON OF TIME COMPLEXITIES

As in the last section, let $\Phi \circ \Psi$ be the map calculated by Algorithm 1 Fill-in(G, π), followed by Algorithm 2 GeneralTreeDecomposition(G, π).

According to the diagram (6.2), a tree decomposition $(\mathbf{T}', \mathcal{X}')$ of the graph of the form $\mathcal{G} + e$, can be done either through the transformation $B \circ \Phi \circ \Psi$ or through $\Phi \circ \Psi \circ A$.

If the edge e maintains the triangulation $\mathbf{H} = \Psi(\mathbf{G})$, then $\mathbf{T}' = \mathbf{T}$, hence $B \circ \Phi \circ \Psi(\mathbf{G}) = \Phi \circ \Psi(\mathbf{G})$. By Remark 6.8 the checking of whether e maintains the triangulation is proportional to the square of the degree of the triangulation. Thus, in general the time complexity of $B \circ \Phi \circ \Psi(\mathbf{G})$ is $O(n^2 m)$, where $n = |V(\mathbf{G})|$ and $m = |E(\mathbf{G})|$.

On the other hand, the processing of $\Phi \circ \Psi \circ A$ entails a time complexity $O(n^2(m + 1))$, since a new edge e is added, and also the checking of whether e maintains the triangulation is performed.

Thus, we see that the processing of $B \circ \Phi \circ \Psi$ is more convenient than that of $\Phi \circ \Psi \circ A$.

This work can be generalized as in [18, 19] in order to provide a deterministic algorithm for tree decompositions for a greater graph class. The generalization entails a potential application in the solution of graph problems with the use of dynamic programming approaches.

6.3   A STRATEGY TO SOLVE NP-HARD PROBLEMS

In this section a brief introduction to the applications of the tree decomposition of graphs is given. We show how to exploit a tree decomposition to solve optimization problems using a dynamic programming approach. We use the language of MSOL to express properties of graph problems. Finally, the *Courcelle theorem* is introduced and a possible application in the design of local Hamiltonian operators for AQC is given.

6.3.1   DYNAMIC PROGRAMMING APPROACH

Let $k \in \mathbb{Z}^+$ be an integer. Let $\mathcal{B}_k$ be the class of graphs $\mathbf{G} = (V, E)$ such that $tw(\mathbf{G}) \le k$.

Theorem 6.15 [81], [82]   For each $k \in \mathbb{Z}^+$, $\mathcal{B}_k$ can be characterized by finite sets of forbidden minors.

The dynamic programming approach for solving NP-hard problems consists in solving partial instances and then to ensemble the corresponding solutions into a solution of the whole initial instance. The deal is *computing tables of characterizations of partial solutions.*

A *nice tree decomposition* $(\mathbf{T} = (T, F), \mathcal{X})$ is a rooted binary tree having nodes of just four types:

- a *start*, a node with no children;

- a *join*, a node with two children, and whose bag is the union of its children's bags;

- a *forget*, a node with just one node, and whose bag is a subset of its child bag; and

- an *introduce*, a node with just one node, and whose bag is a superset of its child bag.

Proposition 6.16 [59] Any tree decomposition of width k of a graph **G** can be transformed into a nice decomposition tree of the same width with $O(k \, \text{card} \, (V(\mathbf{G})))$ nodes in linear time.

Thus, from now on, it can be assumed that all tree decompositions are nice.

A *partial solution* of a problem corresponds to a bag in the decomposition tree. The computation of partial solutions is performed bottom-up ("bottom" corresponds to the leaves, "up" to the root), thus the partial solution at any node is computed from the partial solutions of its children. The partial solution at the root will be the *whole solution*.

For any bag X_t with $t \in T$ in the tree decomposition, let $\mathbf{G}_t$ be the subgraph of **G** whose nodes are the vertexes at the bag X_t and its descendants:

$$V(\mathbf{G}_t) = \bigcup \{t' \in T \mid X_t = X_{t'} \text{ or } [t' \text{ is a descendant of } t \text{ in } \mathbf{T}]\}.$$

In a general way, the following procedural steps synthesizes the dynamic programming reduction: Let P be a problem that given a graph **G** has associated a solution sol$_P$(**G**).

1. Define the notion of a *partial solution*. For a bag X_t for some $t \in T$, it should be the restriction to $\mathbf{G}_t$ of a solution sol$_P$(**G**). Observe that this may coincide or not with sol$_P$($\mathbf{G}_t$).

2. Define the notion of *partial solutions extension* within a tree decomposition.

3. Define the notion of *partial solution characteristic* within a tree decomposition. Most generally, the characteristic of a partial solution at a bag X_t for some $t \in T$ is the restriction to the bag X_t of the partial solution at the graph $\mathbf{G}_t$.

4. Show that for any of the three bag types, there is a polynomial-time algorithm to find the characteristic.

5. Show that the characteristic of the root produces indeed a whole solution.

6.3.2 THE COURCELLE THEOREM

Courcelle showed that every problem definable in Monadic Second-Order Logic (MSOL) can be solved in linear time on graphs with bounded treewidth. The Courcelle theorem has important applications for several fixed parameter tractability results. Many problems can be expressed in MSOL such as Minimum Vertex Cover, Minimum Dominating Set, and Maximum Independent.

Using the definitions of the previous section on tree decompositions, let

$$\mathcal{B}_k = \{\mathbf{G} \mid \mathbf{G} \text{ a graph and tw}(\mathbf{G}) \le k\}.$$

We say that if φ is a well-formed sentence in MSOL for a graph **G** then **G** is called a model for φ.

Theorem 6.17 Courcelle Theorem [27] Let $k \ge 1$ and φ a sentence in MSOL. There exists a linear time algorithm such that for each graph $\mathbf{G} \in \mathcal{B}_k$ decide if **G** is a model of φ.

An application of the Courcelle theorem   A *kernel* in a directed graph $\mathbf{G} = (V, E)$ is a subset K of V such that, no two vertices in K are adjacent and for every vertex $a \in V \setminus K$ there is a vertex $b \in K$ such that $(a, b) \in E$.

Let $k \geq 1$ and $\mathbf{G}$ be a graph with $\mathrm{tw}(\mathbf{G}) \leq k$. There exist a liner time algorithm which decide whether $\mathbf{G}$ has a kernel.

6.3.3   EXAMPLES OF SECOND-ORDER FORMULAE

Independent set

An *independent set* U in a graph $\mathbf{G} = (V, E)$ is a set of vertexes, $U \subseteq V$, containing no pair of edge extremes. As a second order formula, this can be stated by:

$$\phi(U, V) \equiv \forall v_0, v_1 \in V : [v_0, v_1 \in U \implies \{v_0, v_1\} \notin E].$$

Three-coloring

For a graph $\mathbf{G} = (V, E)$ a *coloring* with *n-colors* or *n-coloring*, is a map $\gamma_n : V \to N$ where N is a set of cardinality n, such that no pair of adjacent vertexes share a common color:

$$\forall v_0, v_1 \in V : \{v_0, v_1\} \in E \implies \gamma_n(v_0) \neq \gamma_n(v_1).$$

Namely, an *n*-coloring can be realized as the partition $\left(\gamma_n^{-1}(i)\right)_{i \in N}$ of its monochromatic sets, in which all edges traverse them.

In order to put the notion of 3-*colorability* as a second order formula let us introduce the following formulae.

- Set inclusion:
$$\phi_{10}(V, W) \equiv W \subseteq V.$$

- Three sets form a partition of the set of vertexes:
$$\phi_1(V, W_0, W_1, W_2) \equiv \phi_{10}(V, W_0) \wedge \phi_{10}(V, W_1) \wedge \phi_{10}(V, W_2) \wedge$$
$$[\forall v \in V : (v \in W_0) \vee (v \in W_1) \vee (v \in W_2)].$$

- Two vertexes lie in different subsets:
$$\phi_{20}(v_0, v_1, W) \equiv \neg (v_0 \in W \wedge v_1 \in W).$$

- No edge lies within a monochromatic set:
$$\phi_2(V, W_0, W_1, W_2) \equiv \forall v_0, v_1 \in V :$$
$$[\{v_0, v_1\} \in E \implies \phi_{20}(v_0, v_1, W_0) \wedge \phi_{20}(v_0, v_1, W_1) \wedge \phi_{20}(v_0, v_1, W_2)].$$

- 3-colorability is thus stated by the sentence
$$\phi_3(V) \equiv \exists W_0, W_1, W_2 : \phi_1(V, W_0, W_1, W_2) \wedge \phi_2(V, W_0, W_1, W_2).$$

Hamiltonian cycle

A *Hamiltonian cycle* in a graph $G = (V, E)$ is a permutation of the vertexes such that any pair of contiguous vertexes, modulus the order of the graph, form an edge. In order to put it as a second-order formula let us define some predicates:

- Permutation definition: A map is a list of pairs $P = ((i, v_i))_{i=0}^{n-1}$ associating to each $i \in \{0, \ldots, n-1\}$ an unique point $v_i \in V$:

$$\phi_{00}(P, V) \equiv \bigwedge_{i=0}^{n-1} [\forall v_0, v_1 \in V : [(i, v_0) \in P \wedge (i, v_1) \in P \implies v_0 = v_1]],$$

and P is a permutation if it is one-to-one:

$$\phi_{01}(P, V) \equiv \bigwedge_{0 \leq i < j \leq n-1} [\forall v_0, v_1 \in V : [(i, v_0) \in P \wedge (j, v_1) \in P \implies v_0 \neq v_1]].$$

Let $\phi_0(P, V) = \phi_{00}(P, V) \wedge \phi_{01}(P, V)$.

- Each vertex has an index:

$$\phi_1(P, V) \equiv \forall v \in V : \bigvee_{i=0}^{n-1} [(i, v_0) \in P].$$

- Two indexes correspond to an edge:

$$\phi_{2ij}(P, V) \equiv \forall v_0, v_1 \in V : [(i, v_0) \in P \wedge (j, v_1) \in P \implies \{v_0, v_1\} \in E].$$

- Hamiltonian cycle:

$$\phi_3(P, V) \equiv \phi_0(P, V) \wedge \phi_1(P, V) \wedge \left[\bigwedge_{i=0}^{n-2} \phi_{2,i,i+1}(P, V) \right] \wedge \phi_{2,n-1,0}(P, V).$$

Vertex multicut

For an undirected graph $G = (V, E)$ and a set of pairs $H \subseteq V \times V$, a *vertex multicut* in G consists of a subset $V' \subseteq V$ whose removal from G separates each pair of vertices in H, i.e., there not exists a path in $G - V'$ that connects vertices $(s_j, t_j) \in H$ for every $j = 0, \ldots, m-1$ where $m = \text{card}(H)$. The *Minimal Vertex Multicut Problem* (MVMC) consists in finding a vertex multicut of minimal cardinality. It has been proved that the MVMC is NP-complete for general graphs [22, 34] and remain hard even for graphs with bounded treewidth [22, 41]

On a structure $\lfloor G \rfloor$ let $\phi_{connect}(S, x, y)$ be defined as follows:

$$\phi_{connect}(S, x, y) \;\equiv\; S(x) \wedge S(y) \wedge \forall P\big((P(x) \wedge \neg P(y)) \\ \rightarrow (\exists v \exists w (S(v) \wedge S(w) \wedge P(v) \\ \wedge \neg P(w) \wedge E(v, w)))\big).$$

The formula $\phi_{connect}(S, x, y)$ states that there is a path in $\mathbf{G}$ that connects vertex x to vertex y, and this path lies entirely in $S \subseteq V$, where P is a monadic predicate identified with a partition $\{P_0, P_1\}$ of V such that P_0 is a set of all vertices where P is true and P_1 is a set of all vertices where P is false.

Given a set of pairs $H \subseteq V \times V$ on $\mathbf{G} = (V, E)$, a subset $X \subseteq V$ is a vertex multicut if the following expression is true:

$$\phi_{vmc}(X) \;\equiv\; \forall x \forall y \big(H(x, y) \rightarrow \forall S (\text{connect}(S, x, y) \\ \rightarrow \exists v (X(v) \wedge S(v)))\big)$$

In other words, the formula $\phi_{vmc}(X)$ defines X to be such that for each pair $(x, y) \in H$, whenever there is a subset $S \subseteq V$ that contains a path from x to y, then X must intersect S, i.e., must contain some vertex from S (see [44] for a proof).

6.3.4 DYNAMIC PROGRAMMING APPLIED TO NP-HARD PROBLEMS

The dynamic programming approach for solving NP-hard problems consists of solving partial instances and then ensembling the corresponding solutions into a solution of the whole initial instance.

A *partial solution* of a problem corresponds to a bag in the tree decomposition. The computation of partial solutions is performed bottom-up ("bottom" corresponds to the leaves, "up" to the root), thus the partial solution at any node is computed from the partial solutions of its children. The partial solution at the root will be the *whole solution*.

In the following we expand the procedural steps of the dynamic programming approach sketched in Section 6.3.1, which we based on [15].

A *terminal graph* is a triple $\mathbf{H} = (V, E, X)$ where (V, E) is a graph and the elements of $X \subseteq V$ are called the *terminals* of (V, E). Let $\mathbf{H}_1$ and $\mathbf{H}_2$ be two terminal graphs, $\mathbf{H}_1 \oplus \mathbf{H}_2$ is the disjoint union of $\mathbf{H}_1$ and $\mathbf{H}_2$. A terminal graph $\mathbf{H}_1$ is a terminal subgraph of a graph $\mathbf{G}$ if and only if, there exists a terminal graph $\mathbf{H}_2$ such that $\mathbf{G} = \mathbf{H}_1 \oplus \mathbf{H}_2$.

1. Define a notion of *solution*. Let Π be a problem or a graph property. Let $\text{sol}_\Pi(\mathbf{G}, s)$ be a formula with two variables with $\mathbf{G}$ a graph and s a solution for the instance problem $\mathbf{G}$, such that:

$$\Pi(\mathbf{G}) \iff \exists s : \text{sol}_\Pi(\mathbf{G}, s).$$

2. Define the notion of *partial solution*: A partial solution is an object associated with a terminal graph. Let $\text{psol}_\Pi(\mathbf{H}, s)$ be a formula with two variables with $\mathbf{H}$ a terminal graph and s a partial solution.

3. Define a notion of *extension of partial solutions*. Let $\text{ex}_\Pi(\mathbf{G}, s, \mathbf{H}, s')$ be a formula with four variables with $\mathbf{G}$ a graph, s a solution for the instance problem $\mathbf{G}$, $\mathbf{H}$ a terminal graph and s' a terminal solution for H. The following must hold, for all $\mathbf{G}, s, \mathbf{H}, s'$:

$$\text{ex}_\Pi(\mathbf{G}, s, \mathbf{H}, s') \implies \exists \mathbf{H}' : \mathbf{G} = \mathbf{H} \oplus \mathbf{H}' \wedge \text{sol}_\Pi(\mathbf{G}, s) \wedge \text{sol}_\Pi(\mathbf{H}, s').$$

The following condition expresses that every solution has a partial solution on any terminal graph:

$$\forall \mathbf{G}, s, \mathbf{H}, \mathbf{H}' : (\text{sol}_\Pi(\mathbf{G}, s) \wedge \mathbf{G} = \mathbf{H} \oplus \mathbf{H}') \implies \exists s' : \text{psol}_\Pi(\mathbf{H}, s') \wedge \text{ex}_\Pi(\mathbf{G}, s, \mathbf{H}, s').$$

4. Define a notion of *characteristic of a partial solution*. It is meant to describe what is needed to know about the partial solution to see whether it can be extended to a solution. Let $\text{ch}_\Pi(\mathbf{H}, s) \mapsto \text{psol}_\Pi(\mathbf{H}, s)$ be a function with $\mathbf{H}$ a terminal graph and s a terminal solution for $\mathbf{H}$. It must fulfill, for all terminal graphs $\mathbf{H}, \mathbf{H}', \mathbf{H}''$ and terminal solutions s, s':

$$(\text{ch}_\Pi(\mathbf{H}, s) = \text{ch}_\Pi(\mathbf{H}', s')) \implies (\exists s :'' \text{ex}_\Pi(\mathbf{H} \oplus \mathbf{H},'' s,'' \mathbf{H}, s)) \iff$$
$$(\exists s' :'' \text{ex}_\Pi(\mathbf{H}' \oplus \mathbf{H},'' s',\!'' \mathbf{H}', s')).$$

5. A *full set of characteristics* for a terminal graph $\mathbf{G}$ is the set of all characteristics of partial solutions. For instance, let $\mathbf{H}$ be a terminal graph, the full set of characteristics for $\mathbf{H}$ is:

$$\text{full}_\Pi(\mathbf{H}) = \{\text{ch}_\Pi(\mathbf{H}, s) \mid \text{psol}_\Pi(\mathbf{H}, s)\}.$$

Show that for every type of nodes in a nice tree decomposition, there is a polynomial-time algorithm to find the full set of characteristics.

6. Show that the characteristic of the root produces indeed a whole solution.

A solution to the Maximum Weight Independent Set problem.

Let $\mathbf{G} = (V, E, c)$ be a *weighted graph* where $c : V \to \mathbb{Z}^+$ is a weighted map. An *independent set* is a subset $S \subseteq V$ such that $\forall u, v \in S : \{u, v\} \notin E$. For any subset $S \subseteq V$, the *cost* of S is defined as $c(S) = \sum_{v \in S} c(v)$.

The Maximum Weight Independent Set problem
Input: A weighted graph $\mathbf{G} = (V, E, c)$
Solution: An independent set $S \subseteq V$ with maximum cost.

Let $(\mathbf{T} = (T, F), \mathcal{X})$ be a nice tree decomposition of $\mathbf{G}$. For each $t \in T$, let $\mathbf{G}_t = (V_t, E_t)$ be the subgraph of $\mathbf{G}$ at node t where $V_t = \{v | v \in X_{t_1} \,\&\, (t_1 = t \text{ or } t_1 \text{ is a descendant of } t \text{ in } \mathbf{T})\}$ and $E_t = \{\{u, v\} \in E | u, v \in V_t\}$. The table of characteristics C_t at node t is such that, $\forall S \subseteq X_t, C_t(S) = \max_{W \subseteq V_t}\{c(W) | X_t \cap W = S \,\&\, W \text{ is an independent set}\}$, in case that such an independent set does not exist $C_t(S) = -\infty$.

The procedural steps of the dynamic programming approach will compute the tables C_t for all $t \in T$ in a bottom-up order. We proceed by cases according to the types of nodes on $\mathbf{T}$:

- *Start.* If t is a leaf node of $\mathbf{T}$, then $|X_t| = 1$ and $X_t = \{v\}$. The table C_t has only two entries: $C_t(\emptyset) = 0$ and $C_t(\{v\}) = c(v)$.

- *Introduce.* If t is an introduce node with a child t_1, then $X_t = X_{t_1} \cup \{v\}$ for a vertex v. Observe that $\mathbf{G}_t$ is formed from $\mathbf{G}_{t_1}$ by adding v and zero or more edges from v to vertices in X_{t_1}. It is satisfied that v is not adjacent to any vertex in $V_{t_1} - X_{t_1}$. For each $S \subseteq X_{t_1}$ we have the following.

    1. $C_t(S) = C_{t_1}(S)$.

    2. If there is a vertex $w \in S$ with $\{v, w\} \in E$ then $C_t(S \cup \{v\}) = -\infty$.

    3. If for all $w \in S$, $\{v, w\} \notin E$ then $C_t(S \cup \{v\}) = C_{t_1}(S) + c(v)$.

- *Forget.* If t is a forget node with a child t_1, then $X_t = X_{t_1} - \{v\}$ for some vertex v, and the graphs $\mathbf{G}_t, \mathbf{G}_{t_1}$ are the same. Suppose that $v \in X_{t_1} - X_t$ is that vertex.

    For each $S \subseteq X_t : C_t(S) = \max\{C_{t_1}(S), C_{t_1}(S \cup \{v\})\}$.

- *Join.* If t is a join node with children t_1, t_2, then $X_t = X_{t_1} = X_{t_2}$.

    For each $S \subseteq X_t : C_t(S) = C_{t_1}(S) + C_{t_2}(S) - c(S)$.

Lemma 6.18   *The maximum weight of an independent set in* $\mathbf{G}$ *is* $\max_{S \subseteq X_{root}} C_{root}(S)$.

See [15] for a proof.

The independent set with maximum weight given by the dynamic programming solution can be constructed from the characteristic tables and does not introduce additional time complexity.

In the following we propose a solution to the *classical Ising Spin Glass* model based on the dynamic programming approach, we also introduce the Quantum Ising model and its relation with AQC.

6.3.5   THE CLASSICAL ISING MODEL

Let $\mathbf{G} = (V, E)$ be a graph with vertex set V and edge set $E \subset V^{(2)}$. Let $\mathbb{S} = \{-1, +1\}$ be the set of signs. An *assignment* is a map $\sigma : V \to \mathbb{S}$. An *edge weight* map is of the form $e : E \to \mathbb{R}$ and a *vertex weight* map is of the form $w : V \to \mathbb{R}$. In a physical context, an assignment is called a *spin configuration*, a positive edge weight e is said *ferromagnetic* and a negative edge weight is said *antiferromagnetic*. Let us enumerate $V = (v_i)_{i=0}^{n-1}$, thus there are 2^n assignments. For respective edge and vertex weight e, w, let us write $e_{ij} = e(v_i, v_j)$ and $w_i = w(v_i)$. For those weight maps and an assignment σ, their *energy* is

$$\eta(e, w; \sigma) = - \sum_{\{v_i, v_j\} \in E} e_{ij} \sigma(v_i) \sigma(v_j) - \sum_{v_k \in V} w_k \sigma(v_k). \qquad (6.3)$$

An assignment with minimum energy is called a *ground state*.

For a positive constant $\beta > 0$, let us consider the map

$$\phi(e, w, \beta; \cdot) : \sigma \mapsto \phi(e, w, \beta; \sigma) = \exp(-\beta \, \eta(e, w; \sigma)). \qquad (6.4)$$

Let $\Phi(e, w, \beta) = \sum \{\phi(e, w, \beta; \sigma) | \sigma \text{ is an assignment}\}$. Thus, a probability density on the space of assignments is given as

$$\pi(e, w, \beta; \cdot) : \sigma \mapsto \frac{\phi(e, w, \beta; \sigma)}{\Phi(e, w, \beta)}. \qquad (6.5)$$

From relation (6.3) it is evident that if the vertex weight w is null then the energy map is "even:"

$$\forall \sigma : \text{assignment} : \quad \eta(e, 0; \sigma) = \eta(e, 0; -\sigma). \qquad (6.6)$$

For an assignment σ, its *support* is $\text{Spt}(\sigma) = \{v \in V | \sigma(v) = +1\}$. A *2-partition* of V is a collection of the form $\{U, V - U\}$, such that $U \subseteq V$. Clearly $\sigma \leftrightarrow \{\text{Spt}(\sigma), V - \text{Spt}(\sigma)\}$ is a bijective correspondence among assignments and 2-partitions of V.

For any set $U \subseteq V$, let

$$c(U) = \{e \in E | \text{card}(e \cap U) = 1 \ \& \ \text{card}(e \cap (V - U)) = 1\} \qquad (6.7)$$

be the collection of edges with an extreme in U and the other in its complement. Since an assignment is a $\mathbb{S}$-valued map:

$$\forall \sigma : \text{assignment} : \quad \eta(e, 0; \sigma) \quad = \quad - \sum_{\{v_i, v_j\} \in E} e_{ij} + 2 \sum_{\{v_i, v_j\} \in c(\text{Spt}(\sigma))} e_{ij}$$

$$=: \quad \eta_s(e; \text{Spt}(\sigma)). \qquad (6.8)$$

Let us introduce the following problem.

Minimum weight cut
Instance: A graph $\mathbf{G} = (V, E)$ and an edge weighted map $e : E \to \mathbb{R}^+$.
Solution: A partition $\{U, V - U\}$ of the vertex set V such that $c(U)$, as defined by (6.7), is minimum.

Clearly, this problem is equivalent to minimize the energy operator $\eta(e, 0; \cdot)$ as defined by (6.3), or equivalently to find a vertex set U which minimizes $\eta_s(e; U)$ as defined by (6.8).

A similar problem is the following.

Two-dimensional magnetic field
Instance: A planar graph $\mathbf{G} = (V, E)$.
Solution: A ground state σ_0 for the operator

$$\sigma \mapsto \eta(-1, -1; \sigma) = \sum_{\{v_i, v_j\} \in E} \sigma(v_i) \sigma(v_j) + \sum_{v_k \in V} \sigma(v_k). \tag{6.9}$$

Theorem 6.19 Barahona [11] *Two-dimensional magnetic field* is an NP-complete problem.

For edge and vertex weight e and w, we refer to the Ising spin model as the map:

$$\forall \sigma \in \text{assignment} : \eta(-e, -w; \sigma) = \sum_{\{v_i, v_j\} \in E} e_{ij} \sigma(v_i) \sigma(v_j) + \sum_{v_k \in V} w_k \sigma(v_k) \tag{6.10}$$

and for any subset $S \subseteq V$, let

$$\eta(-e, -w; \sigma, S) = \sum_{\{v_i, v_j\} \in E(\mathbf{G}[S])} e_{ij} \sigma(v_i) \sigma(v_j) + \sum_{v_k \in S} w_k \sigma(v_k)$$

be the Ising spin model defined on $\mathbf{G}[S]$.

The Ising spin model can be solved using the dynamic programming approach by solving partial solutions at every node on a tree-decomposition, and by the Courcelle Theorem there exists an algorithm with polynomial time complexity that solve the Ising spin glass model over graph instances with bounded treewidth.

The dynamic programming solution is as follows.

Let $\mathbf{G} = (V, E)$ be a graph with edge and vertex weights e and w. Let $(\mathbf{T} = (T, F), \mathcal{X})$ be a nice tree decomposition of $\mathbf{G}$ and assume that the treewidth of $(\mathbf{T}, \mathcal{X})$ is bounded by a constant k. We compute the tables C_t for all $t \in T$ in a bottom-up order. We proceed by cases according to the types of nodes on $\mathbf{T}$.

- *Start.* If t is a leaf node of $\mathbf{T}$, then $|X_t| = 1$ and $X_t = \{v_i\}$ for $v_i \in V$. The table C_t has only two entries: $\forall \tau \in \{0, 1\} : C_t(\{v_i\}, \tau) = \eta(-e, -w; \sigma(v_i) = \tau, \{v_i\})$.

- *Introduce.* If t is an introduce node with a child t_1, then $X_t = X_{t_1} \cup \{v_i\}$ for $v_i \in V$. For all $\tau \in \{0, 1\}^{|X_t| - 1}$:

  1. for all $v_j \in X_{t_1}, b \in \{0, 1\}$: if $\{v_i, v_j\} \in E$ then $C_t(X_t, \tau \cdot b) = C_{t_1}(X_{t_1}, \tau) + e_{ij} \sigma(v_i) \sigma(v_j) + w_i \sigma(v_i)$, otherwise $C_t(X_t, \tau \cdot b) = C_{t_1}(X_{t_1}, \tau) + w_i \sigma(v_i)$.

- *Forget.* If t is a forget node with a child t_1, then $X_t = X_{t_1} - \{v_i\}$ for $v_i \in V$. For all $\tau \in \{0,1\}^{|X_t|}$:

  1. $C_t(X_t, \tau) = \min\{C_t(X_{t_1}, \tau \cdot 0), C_t(X_{t_1}, \tau \cdot 1)\}$.

- *Join.* If t is a join node with children t_1, t_2 then $X_t = X_{t_1} = X_{t_2}$. For all $\tau \in \{0,1\}^{|X_t|}$:
  $$C_t(X_t, \tau) = C_{t_1}(X_{t_1}, \tau) + C_{t_2}(X_{t_2}, \tau) - \eta(-e, -w; \sigma(X_t) = \tau, \{X_t\}).$$

It is easy to prove that the energy ground state of **G** can be obtained from the table C_{root}. A similar algorithm for the Ising spin model is the proposed in [10], where a conditional restriction is imposed on the set of nodes of the tree-decomposition.

6.3.6 QUANTUM ISING MODEL

Let us consider the Pauli transforms $\sigma_x, \sigma_z : \mathbb{H}_1 \to \mathbb{H}_1$ whose matrices with respect to the canonical basis are

$$\sigma_x = \begin{pmatrix} 0 & 1 \\ 1 & 0 \end{pmatrix} \quad , \quad \sigma_z = \begin{pmatrix} 1 & 0 \\ 0 & -1 \end{pmatrix}. \tag{6.11}$$

Over the canonical basis, we have

$$\forall \varepsilon \in Q : \quad \sigma_z(|\varepsilon\rangle) = \theta(\varepsilon) \, |\varepsilon\rangle, \tag{6.12}$$

where $\theta : Q \to \mathbb{S}, \varepsilon \mapsto 1 - 2\varepsilon$.

Let $n \in \mathbb{Z}^+$ be a positive integer and let $[\![0, n-1]\!] = \{0, \dots, n-1\}$ be the initial segment of the natural numbers with n elements.

For any index $j \in [\![0, n-1]\!]$ let $\sigma_z^j = \bigotimes_{v=0}^{n-1} s_v : \mathbb{H}_n \to \mathbb{H}_n$, where $s_v = \sigma_z$ if $v = j$ and $s_v = \mathrm{Id}$ otherwise. In other words, σ_z^j applies the transform σ_z at the j-th qubit of any n-quregister in $\mathbb{H}_n$. Then, as in (6.12):

$$\forall j \in [\![0, n-1]\!], \ \varepsilon \in Q^n : \quad \sigma_z^j(|\varepsilon\rangle) = \theta(\varepsilon_j) \, |\varepsilon\rangle, \tag{6.13}$$

Let $\mathbf{G} = (V, E)$ be a graph whose vertices are the first n indexes, $V = [\![0, n-1]\!]$, and $E \subseteq [\![0, n-1]\!]^{(2)}$ is a set of index pairs.

For any vertex weighting map $w : V \to \mathbb{R}$ let us consider the operator

$$H_w : \mathbb{H}_n \to \mathbb{H}_n \, , \quad H_w = \sum_{j=0}^{n-1} w_j \sigma_z^j. \tag{6.14}$$

From (6.13) we have

$$\forall \varepsilon \in Q^n : \quad H_w(|\varepsilon\rangle) = \left(\sum_{j=0}^{n-1} w_j \theta(\varepsilon_j) \right) |\varepsilon\rangle, \tag{6.15}$$

hence H_w is a diagonal operator.

Similarly, for any edge weighting map $e : E \to \mathbb{R}$, let us consider the operator

$$H_e : \mathbb{H}_n \to \mathbb{H}_n , \quad H_e = \sum_{\{i,j\} \in E} e_{ij} \, \sigma_z^i \circ \sigma_z^j. \tag{6.16}$$

Since the operators σ_z^j are pairwise commutative, again from (6.13) we have

$$\forall \varepsilon \in Q^n : \quad H_e(|\varepsilon\rangle) = \left(\sum_{\{i,j\} \in E} e_{ij} \, \theta(\varepsilon_i) \theta(\varepsilon_j) \right) |\varepsilon\rangle ; \tag{6.17}$$

hence, H_e is as well a diagonal operator.

As in Equation (6.3) let us define the operator

$$H(e, w; \cdot) : \mathbb{H}_n \to \mathbb{H}_n , \quad H(e, w; \cdot) = -H_e - H_w. \tag{6.18}$$

From (6.15) and (6.17) we have

$$\forall \varepsilon \in Q^n : \quad H(e, w; |\varepsilon\rangle) = \left(- \sum_{\{i,j\} \in E} e_{ij} \, \theta(\varepsilon_i) \theta(\varepsilon_j) - \sum_{j=0}^{n-1} w_j \theta(\varepsilon_j) \right) |\varepsilon\rangle . \tag{6.19}$$

Between the greatest parenthesis, an energy map $\eta(e, w; \cdot) : Q^n \to \mathbb{R}$ appears of the type of Equation (6.3), and a ground state $|\varepsilon_0\rangle$ of H corresponds naturally with a ground state ε_0 of $\eta(e, w; \cdot)$.

On the other hand, the Pauli transform σ_x (see (6.11)) has eigenvalues $+1, -1$ with respective eigenvectors $c_0 = W |0\rangle$ and $c_1 = W |1\rangle$, where W is the Hadamard transform. For any index $j \in [\![0, n-1]\!]$ let $\sigma_x^j = \bigotimes_{\nu=0}^{n-1} r_\nu : \mathbb{H}_n \to \mathbb{H}_n$, where $r_\nu = \sigma_x$ if $\nu = j$ and $r_\nu = \mathrm{Id}$ otherwise. In other words, σ_x^j applies the transform σ_x at the j-th qubit of any n-quregister in $\mathbb{H}_n$. Then, as in (6.12):

$$\forall j \in [\![0, n-1]\!], \ \varepsilon \in Q^n : \quad \sigma_x^j(c_\varepsilon) = \theta(\varepsilon_j) \, c_\varepsilon, \tag{6.20}$$

where $c_\varepsilon = \bigotimes_{i=0}^{n-1} c_{\varepsilon_i}$.

For any vertex weighting map $h : V \to \mathbb{R}$ let us introduce the operator

$$H_h : \mathbb{H}_n \to \mathbb{H}_n , \quad H_h = \sum_{j=0}^{n-1} h_j \sigma_x^j. \tag{}$$

From (6.20) we have

$$\forall \varepsilon \in Q^n : \quad H_h(c_\varepsilon) = \left(\sum_{j=0}^{n-1} h_j \theta(\varepsilon_j) \right) c_\varepsilon, \tag{6.21}$$

hence, H_d is a diagonal operator, and a ground state has the form c_{ε_0} for $\varepsilon_0 \in Q^n$ minimizing $\sum_{j=0}^{n-1} h_j \theta(\varepsilon_j)$, which in turn can easily be calculated depending on the map h.

Thus, the problem to find a ground state of the operator $H(e, w; \cdot)$ determined by (6.19) can be solved using the Adiabatic Theorem with the operator path:

$$H_t = \left(1 - \frac{t}{T}\right) H_h + \frac{t}{T} H(e, w; \cdot)$$

for some large enough $T \in \mathbb{R}^+$.

CHAPTER 7

Conclusions

We investigated the application of the adiabatic quantum computing to solve NP-hard problems. We showed a procedural construction of the Hamiltonian operators for adiabatic computing for the MAX-SAT problem. This construction can be extended to describe the Hamiltonian operators of other NP-hard problems with similar structure.

We investigated the construction of local Hamiltonian operators for Adiabatic Quantum Computing. It is based on the dynamic programming approach and the monadic second order logic. We showed results to modify an initial tree decomposition when new edges to the input graph are added. We showed a general methodology to construct local Hamiltonian operators based on the Ising model in which the energy function minimization is equivalent to the optimization problem of pseudo Boolean functions. The Ising Hamiltonians have been used to design local Hamiltonian for the Adiabatic Quantum Computing (AQC), and they have a natural use to describe graph problems and other optimization problems.

We investigated the properties of the optimization problem of pseudo Boolean functions to construct local Hamiltonian operators. We considered the quadratic optimization problem, since every pseudo boolean optimization problem is equivalent to the quadratic case. Hence, at the present time we are involved in the decomposition of initial and ending Hamiltonians, arisen from the AQC approach, into a sum of local Hamiltonians.

The future work of this thesis is the following: to improve the symbolic characterization provided in Chapter 4 of the AQC solution to many others combinatorial optimization problems. It is important to have a general characterization of the initial and final Hamiltonians for any NP-hard problems. In Chapter 6, we showed a general methodology to construct AQC algorithms, as well as demonstrating the fact that every MSOL expression has associated quadratic pseudo-Boolean forms. It is important to classify optimization problems in terms of its representability, for instance according to the polynomial hierarchy.

The dynamic programming approach studied in Chapter 6 was considered in [10]. This approach depends on the possible physical implementation of quantum memories; the dynamic programming solution requires the storing of partial solutions. This still remains as an open problem in the design of quantum algorithms.

Finally, a new tool in the design of quantum algorithms is the geometric Berry phase; these kinds of algorithms are robust to some sources of errors. The Berry phase can encode the solution of search problems. It is important to find a general technique to encode the solution of optimization problems into the Berry phase.

Bibliography

[1] Dorit Aharonov and Amnon Ta-Shma. Adiabatic quantum state generation. *SIAM Journal on Computing*, 37:47–82, April 2007. DOI: 10.1137/060648829. 2

[2] Dorit Aharonov, Wim van Dam, Julia Kempe, Zeph Landau, Seth Lloyd, and Oded Regev. Adiabatic quantum computation is equivalent to standard quantum computation. *SIAM Journal on Computing*, 37:166–194, May 2007. DOI: 10.1137/S0097539705447323. 1, 24

[3] Noga Alon, Laszlo Babai, and Alon Itai. A fast and simple randomized parallel algorithm for the maximal independent set problem. *Journal of Algorithms*, 7(4):567 – 583, 1986. DOI: 10.1016/0196-6774(86)90019-2. 17

[4] Boris Altshuler, Hari Krovi, and Jérémie Roland. Adiabatic quantum optimization fails for random instances of NP-complete problems. *CoRR*, abs/0908.2782, 2009. 2

[5] Stefan Arnborg. Decomposable structures, boolean function representations, and optimization. In *Proceedings of the 20th International Symposium on Mathematical Foundations of Computer Science*, MFCS '95, pages 21–36, London, UK, UK, 1995. Springer-Verlag. DOI: 10.1007/3-540-60246-1_110. 63

[6] Stefan Arnborg, Derek G. Corneil, and Andrzej Proskurowski. Complexity of finding embeddings in a k-tree. *SIAM Journal on Algebraic Discrete Methods*, 8(2):277–284, 1987. DOI: 10.1137/0608024. 64

[7] Sanjeev Arora and Boaz Barak. *Computational Complexity: A Modern Approach*. Cambridge University Press, New York, NY, USA, 1st edition, 2009. DOI: 10.1017/CBO9780511804090. 5

[8] Sanjeev Arora and Shmuel Safra. Probabilistic checking of proofs: a new characterization of np. *Journal of the ACM*, 45:70–122, January 1998. DOI: 10.1145/273865.273901. 7

[9] Giorgio Ausiello, G. Gambosi, P. Crescenzi, V. Kann, A. Marchetti-Spaccamela, and M. Protasi. *Complexity and Approximation: Combinatorial Optimization Problems and Their Approximability Properties*. Springer, 1999. 5, 8, 63

[10] Nikhil Bansal, Sergey Bravyi, and Barbara M. Terhal. Classical approximation schemes for the ground-state energy of quantum and classical ising spin hamiltonians on planar graphs. *Quantum Information & Computation*, 9(7):701–720, 2009. 2, 81, 85

[11] Francisco Barahona. On the computational complexity of ising spin glass models. *Journal of Physics A: Mathematical and General*, 15(10):3241–3253, 1982. DOI: 10.1088/0305-4470/15/10/028. 47, 80

[12] Ethan Bernstein and Umesh Vazirani. Quantum complexity theory. *SIAM Journal on Computing*, 26:11–20, 1997. DOI: 10.1137/S0097539796300921. 12

[13] D. W. Berry, G. Ahokas, R. Cleve, and B. C. Sanders. Efficient Quantum Algorithms for Simulating Sparse Hamiltonians. *Communications in Mathematical Physics*, 270:359–371, March 2007. DOI: 10.1007/s00220-006-0150-x. 2

[14] Hans Bodlaender. Treewidth: Algorithmic techniques and results. In Igor Prívara and Peter Ružička, editors, *Mathematical Foundations of Computer Science 1997*, volume 1295 of *Lecture Notes in Computer Science*, pages 19–36. Springer Berlin/Heidelberg, 1997. 10.1007/BFb0029946. DOI: 10.1007/BFb0029943. 2, 3, 63

[15] Hans L. Bodlaender and Arie M. C. A. Koster. Combinatorial optimization on graphs of bounded treewidth. *The Computer Journal*, 51(3):255–269, 2008. DOI: 10.1093/comjnl/bxm037. 76, 78

[16] Hans L. Bodlaender and Arie M. C. A. Koster. Treewidth computations II. lower bounds. *Information and Computation*, 209(7):1103–1119, 2011. DOI: 10.1016/j.ic.2011.04.003.

[17] Hans L. Bodlaender and Arie M.C.A. Koster. Treewidth computations I. upper bounds. *Information and Computation*, 208(3):259 – 275, 2010. DOI: 10.1016/j.ic.2009.03.008. 2, 3, 63

[18] R. B. Borie. Generation of polynomial-time algorithms for some optimization problems on tree-decomposable graphs. *Algorithmica*, 14:123–137, 1995. 10.1007/BF01293664. DOI: 10.1007/BF01293664. 2, 72

[19] Richard B. Borie, R. Gary Parker, and Craig A. Tovey. Deterministic decomposition of recursive graph classes. *SIAM Journal on Discrete Mathematics*, 4:481–501, September 1991. DOI: 10.1137/0404043. 72

[20] Endre Boros and Peter L. Hammer. Pseudo-boolean optimization. *Discrete Applied Mathematics*, 123:155–225, November 2002. DOI: 10.1016/S0166-218X(01)00341-9. 2, 43, 44, 55

[21] Sergey Bravyi, David P. DiVincenzo, Roberto Oliveira, and Barbara M. Terhal. The complexity of stoquastic local hamiltonian problems. *Quantum Information & Computation*, 8(5):361–385, 2008. 2

[22] Gruia Calinescu, Cristina G. Fernandes, and Bruce Reed. Multicuts in unweighted graphs and digraphs with bounded degree and bounded tree-width. *Journal of Algorithms*, 48(2):333–359, September 2003. DOI: 10.1016/S0196-6774(03)00073-7. 75

[23] A. M. Childs. *Quantum information processing in continuous time. PhD thesis.* Massachusetts Institute of Technology, 2004. 2

[24] Vicky Choi. Minor-embedding in adiabatic quantum computation: I. The parameter setting problem. *Quantum Information Processing*, 7:193–209, 2008. 10.1007/s11128-008-0082-9. DOI: 10.1007/s11128-008-0082-9. 47

[25] Vicky Choi. Different adiabatic quantum optimization algorithms. *Quantum Information & Computation*, 11(7&8):638–648, 2011. DOI: 10.1073/pnas.1018310108. 2

[26] Vicky Choi. Minor-embedding in adiabatic quantum computation: Ii. minor-universal graph design. *Quantum Information Processing*, 10:343–353, 2011. 10.1007/s11128-010-0200-3. DOI: 10.1007/s11128-010-0200-3. 47

[27] Bruno Courcelle. *Graph Structure and Monadic Second-Order Logic.* Cambridge University Press, 2011. 55, 63, 73

[28] Bruno Courcelle, Johann A. Makowsky, and Udi Rotics. Linear time solvable optimization problems on graphs of bounded clique-width. *Theory Computing Systems*, 33(2):125–150, 2000. DOI: 10.1007/s002249910009. 55, 57, 63

[29] Bruno Courcelle, Johann A. Makowsky, and Udi Rotics. On the fixed parameter complexity of graph enumeration problems definable in monadic second-order logic. *Discrete Applied Mathematics*, 108(1-2):23–52, 2001. DOI: 10.1016/S0166-218X(00)00221-3. 63

[30] Bruno Courcelle and Stephan Olariu. Upper bounds to the clique width of graphs. *Discrete Applied Mathematics*, 101(1-3):77–114, April 2000. DOI: 10.1016/S0166-218X(99)00184-5. 63

[31] Yves Crama and Peter L. Hammer. *Boolean Functions - Theory, Algorithms, and Applications*, volume 142 of *Encyclopedia of mathematics and its applications.* Cambridge University Press, 2011. DOI: 10.1017/CBO9780511852008. 59

[32] William Cruz-Santos and Guillermo Morales-Luna. On the hamiltonian operators for adiabatic quantum reduction of sat. In *LATA*, pages 239–248, 2010. DOI: 10.1007/978-3-642-13089-2_20. 31

[33] William Cruz-Santos and Guillermo Morales-Luna. Guided evolution of tree decompositions of graphs. *International Journal of Computational and Applied Mathematics*, 7(1):13–24, 2012. 63

[34] E. Dahlhaus, D. S. Johnson, C. H. Papadimitriou, P. D. Seymour, and M. Yannakakis. The complexity of multiterminal cuts. *SIAM Journal on Computing*, 23(4):864–894, August 1994. DOI: 10.1137/S0097539792225297. 75

[35] L.-M. Duan, J. I. Cirac, and P. Zoller. Geometric Manipulation of Trapped Ions for Quantum Computation. *Science*, 292:1695–1697, 2001. DOI: 10.1126/science.1058835. 30

[36] Edward Farhi, Jeffrey Goldstone, Sam Gutmann, Joshua Lapan, Andrew Lundgren, and Daniel Preda. A quantum adiabatic evolution algorithm applied to random instances of an np-complete problem. *Science*, 292(5516):472–475, 2001. DOI: 10.1126/science.1057726. 2

[37] Edward Farhi, Jeffrey Goldstone, Sam Gutmann, and Daniel Nagaj. How to make the quantum adiabatic algorithm fail. *International Journal of Quantum Information*, 06, 2008. DOI: 10.1142/S021974990800358X. 2

[38] Edward Farhi, Jeffrey Goldstone, Sam Gutmann, and Michael Sipser. Quantum computation by adiabatic evolution. arXiv:quant-ph/0001106v1, 2000. 1, 19, 24, 31

[39] Richard P. Feynman. Simulating Physics with Computers. *International Journal of Theoretical Physics*, 21(6/7):467–488, 1982. DOI: 10.1007/BF02650179. 1

[40] D. R. Fulkerson and O. A. Gross. Incidence matrices and interval graphs. *Pacific Journal of Mathematics*, 15(3):835–855, 1965. DOI: 10.2140/pjm.1965.15.835. 64, 65

[41] N. Garg, V.V. Vazirani, and M. Yannakakis. Primal-dual approximation algorithms for integral flow and multicut in trees. *Algorithmica*, 18(1):3–20, 1997. DOI: 10.1007/BF02523685. 75

[42] Oded Goldreich. *Computational Complexity: A Conceptual Perspective.* Cambridge University Press, Published in US in May 2008. DOI: 10.1017/CBO9780511804106. 5, 10, 11, 14, 16

[43] Martin Charles Golumbic and Udi Rotics. On the clique-width of some perfect graph classes. *International Journal of Foundations of Computer Science*, 11(3):423–443, 2000. DOI: 10.1142/S0129054100000260. 63

[44] Georg Gottlob and Stephanie Tien Lee. A logical approach to multicut problems. *Inf. Process. Lett.*, 103(4):136–141, August 2007. DOI: 10.1016/j.ipl.2007.03.005. 55, 76

[45] David J. Griffiths. *Introduction to Quantum Mechanics.* Pearson Prentice Hall, 2005. 1, 19

[46] Lov K. Grover. A fast quantum mechanical algorithm for database search. In *Proceedings of the twenty-eighth annual ACM symposium on Theory of computing*, STOC '96, pages 212–219, New York, NY, USA, 1996. ACM. DOI: 10.1145/237814.237866. 1, 30

[47] Pinar Heggernes. Minimal triangulations of graphs: A survey. *Discrete Mathematics*, 306(3):297 – 317, 2006. Minimal Separation and Minimal Triangulation. DOI: 10.1016/j.disc.2005.12.003. 65

[48] Dorit S. Hochbaum, editor. *Approximation algorithms for NP-hard problems*. PWS Publishing Co., Boston, MA, USA, 1997. 7

[49] Tad Hogg. Adiabatic quantum computing for random satisfiability problems. arXiv:quant-ph/0206059, 2002. 31

[50] Neil Immerman. *Descriptive complexity*. Graduate texts in computer science. Springer, 1999. DOI: 10.1007/978-1-4612-0539-5. 55, 63

[51] Sorin Istrail. Statistical mechanics, three-dimensionality and np-completeness: I. universality of intracatability for the partition function of the ising model across non-planar surfaces (extended abstract). In *Proceedings of the thirty-second annual ACM symposium on Theory of computing*, STOC '00, pages 87–96, New York, NY, USA, 2000. ACM. DOI: 10.1145/335305.335316. 47

[52] David S. Johnson. Approximation algorithms for combinatorial problems. In *Proceedings of the fifth annual ACM symposium on Theory of computing*, STOC '73, pages 38–49, New York, NY, USA, 1973. ACM. DOI: 10.1145/800125.804034. 17

[53] Jonathan A. Jones, Vlatko Vedral, Artur Ekert, and Giuseppe Castagnoli. Geometric quantum computation using nuclear magnetic resonance. *Nature*, 403(6772):869–871, February 2000. DOI: 10.1038/35002720. 30

[54] Richard Jozsa. Classical simulation and complexity of quantum computations. In *CSR*, pages 252–258, 2010. DOI: 10.1007/978-3-642-13182-0_23. 18

[55] Ming-Yang Kao, editor. *Encyclopedia of Algorithms*. Springer, 2008. 63

[56] Julia Kempe, Alexei Kitaev, and Oded Regev. The complexity of the local hamiltonian problem. *SIAM Journal on Computing*, 35(5):1070–1097, 2006. DOI: 10.1137/S0097539704445226. 2, 52

[57] Julia Kempe and Oded Regev. 3-local Hamitonian is QMA-complete. *Quantum Information and Computation*, 3:258–264, May 2003. 52

[58] A. Yu. Kitaev, A. H. Shen, and M. N. Vyalyi. *Classical and Quantum Computation*. American Mathematical Society, Boston, MA, USA, 2002. 2, 51, 52

[59] T. Kloks, H. Bodlaender, H. Müller, and D. Kratsch. Computing treewidth and minimum fill-in: All you need are the minimal separators. In Thomas Lengauer, editor, *Algorithms ESA' 93*, volume 726 of *Lecture Notes in Computer Science*, pages 260–271. Springer Berlin/Heidelberg, 1993. DOI: 10.1007/3-540-57273-2. 65, 73

[60] Ton Kloks. *Treewidth, Computations and Approximations*, volume 842 of *Lecture Notes in Computer Science*. Springer, 1994. 63, 64

[61] P.G. Kolaitis and M.N. Thakur. Logical definability of np optimization problems. *Information and Computation*, 115(2):321 – 353, 1994. DOI: 10.1006/inco.1994.1100. 63

[62] A.E. Margolin, V.I. Strazhev, and A.Ya. Tregubovich. Geometric phases and quantum computations. *Physics Letters A*, 303(2003):131 – 134, 2002. 29

[63] David Marker. *Introduction to model theory*. Graduate texts in mathematics. Springer, 2002. 55

[64] Ernst W. Mayr, Hans Jürgen Prömel, and Angelika Steger, editors. *Lectures on Proof Verification and Approximation Algorithms. (the book grow out of a Dagstuhl Seminar, April 21-25, 1997)*, volume 1367 of *Lecture Notes in Computer Science*. Springer, 1998. 17

[65] Albert Messiah. *Quantum mechanics*. Dover Publications, 1999. 1, 22, 23

[66] Rajeev Motwani and Prabhakar Raghavan. *Randomized algorithms*. Cambridge University Press, New York, NY, USA, 1995. DOI: 10.1017/CBO9780511814075. 18

[67] M. Nakahara. *Geometry, Topology and Physics, Second Edition*. Graduate student series in physics. Taylor & Francis, 2003. DOI: 10.1201/9781420056945. 29

[68] Assaf Natanzon, Ron Shamir, and Roded Sharan. A polynomial approximation algorithm for the minimum fill-in problem. In *Proceedings of the thirtieth annual ACM symposium on Theory of computing*, STOC '98, pages 41–47, New York, NY, USA, 1998. ACM. DOI: 10.1145/276698.276710. 65

[69] J. Nešetřil and P.O. de Mendez. *Sparsity: Graphs, Structures, and Algorithms*. Algorithms and Combinatorics. Springer, 2012. 63

[70] Michael A. Nielsen and Isaac L. Chuang. *Quantum Computation and Quantum Information*. Cambridge University Press, 1 edition, October 2000. 1, 13, 48

[71] Roberto Oliveira and Barbara M. Terhal. The complexity of quantum spin systems on a two-dimensionalkitaev square lattice. *Quantum Information & Computation*, 8(10):900–924, 2010. 2

[72] Ognyan Oreshkov, Todd A. Brun, and Daniel A. Lidar. Fault-tolerant holonomic quantum computation. *Physical Review Letters*, 102:070502, Feb 2009. DOI: 10.1103/PhysRevLett.102.070502. 30

[73] Christos M. Papadimitriou. *Computational complexity*. Addison-Wesley, Reading, Massachusetts, 1994. 55

[74] Anargyros Papageorgiou and Chi Zhang. On the efficiency of quantum algorithms for hamiltonian simulation. *Quantum Information Processing*, 11:541–561, 2012. 10.1007/s11128-011-0263-9. DOI: 10.1007/s11128-011-0263-9. 2

[75] Alejandro Perdomo, Colin Truncik, Ivan Tubert-Brohman, Geordie Rose, and Alán Aspuru-Guzik. Construction of model hamiltonians for adiabatic quantum computation and its application to finding low-energy conformations of lattice protein models. *Physical Review A*, 78(1):012320, Jul 2008. DOI: 10.1103/PhysRevA.78.012320. 2, 47

[76] Alejandro Perdomo-Ortiz, Neil Dickson, Marshall Drew-Brook, Geordie Rose, and Alan Aspuru-Guzik. Finding low-energy conformations of lattice protein models by quantum annealing. *Scientific Reports*, 2:571, 2012. DOI: 10.1038/srep00571. 47

[77] Reinhard Pichler, Stefan Rümmele, and Stefan Woltran. Multicut algorithms via tree decompositions. In Tiziana Calamoneri and Josep Diaz, editors, *Algorithms and Complexity*, volume 6078 of *Lecture Notes in Computer Science*, pages 167–179. Springer Berlin Heidelberg, 2010. DOI: 10.1007/978-3-642-13073-1. 63

[78] Prabhakar Raghavan. Probabilistic construction of deterministic algorithms: approximating packing integer programs. *Journal of Computer and System Sciences*, 37(2):130–143, October 1988. DOI: 10.1016/0022-0000(88)90003-7. 17

[79] N. Robertson and P. D. Seymour. Graph minors. IV. Tree-width and well-quasi-ordering. *Journal of Combinatorial Theory (Series B)*, 48:227–254, April 1990. DOI: 10.1016/0095-8956(90)90120-O. 3, 63

[80] N. Robertson and P. D. Seymour. Graph minors XIII: The disjoint path problem. *Journal of Combinatorial Theory(Series B)*, 63:65–110, 1995. DOI: 10.1006/jctb.1995.1006.

[81] Neil Robertson and P. D. Seymour. Graph minors. II. Algorithmic aspects of tree-width. *Journal of Algorithms*, 7(3):309 – 322, 1986. DOI: 10.1016/0196-6774(86)90023-4. 72

[82] Neil Robertson and P D Seymour. Graph minors. V. Excluding a planar graph. *Journal of Combinatorial Theory (Series B)*, 41:92–114, August 1986. DOI: 10.1016/0095-8956(86)90030-4. 3, 72

[83] Neil Robertson and P. D. Seymour. Graph minors. X. Obstructions to tree-decomposition. *Journal of Combinatorial Theory (Series B)*, 52(2):153 – 190, 1991. DOI: 10.1016/0095-8956(91)90061-N. 63

[84] Donald J. Rose. On simple characterizations of k-trees. *Discrete Mathematics*, 7(3-4):317 – 322, 1974. DOI: 10.1016/0012-365X(74)90042-9. 64

[85] Donald J. Rose and R. Endre Tarjan. Algorithmic aspects of vertex elimination. In *Proceedings of seventh annual ACM symposium on Theory of computing*, STOC '75, pages 245–254, New York, NY, USA, 1975. ACM. DOI: 10.1145/800116.803775. 65

[86] Stuart J. Russell and Peter Norvig. *Artificial Intelligence: A Modern Approach*. Pearson Education, 2003. 59, 61

[87] P. W. Shor. Algorithms for quantum computation: discrete logarithms and factoring. In *Proceedings of the 35th Annual Symposium on Foundations of Computer Science*, pages 124–134, Washington, DC, USA, 1994. IEEE Computer Society. DOI: 10.1109/SFCS.1994.365700. 1

[88] Erik Sjöqvist. A new phase in quantum computation. *Physics*, 1:35, Nov 2008. DOI: 10.1103/Physics.1.35. 29

[89] R. D. Somma, S. Boixo, H. Barnum, and E. Knill. Quantum simulations of classical annealing processes. *Physical Review Letters*, 101:130504, Sep 2008. DOI: 10.1103/PhysRevLett.101.130504. 18

[90] Luca Trevisan. Inapproximability of combinatorial optimization problems. *CoRR*, cs.CC/0409043, 2004. DOI: 10.1002/9781118600207.ch13. 7

[91] Avatar Tulsi. Adiabatic quantum computation with a one-dimensional projector hamiltonian. *Physical Review A*, 80:052328, Nov 2009. DOI: 10.1103/PhysRevA.80.052328. 2

[92] W. van Dam, M. Mosca, and U. Vazirani. How powerful is adiabatic quantum computation? In *Foundations of Computer Science, 2001. Proceedings. 42nd IEEE Symposium on*, pages 279 – 287, oct. 2001. DOI: 10.1109/SFCS.2001.959902. 47

[93] Johan M. M. van Rooij, Hans L. Bodlaender, and Peter Rossmanith. Dynamic programming on tree decompositions using generalised fast subset convolution. In *ESA*, pages 566–577, 2009. DOI: 10.1007/978-3-642-04128-0_51. 63

[94] Vijay V. Vazirani. *Approximation algorithms*. Springer-Verlag New York, Inc., New York, NY, USA, 2001. 7

[95] Pawel Wocjan and Thomas Beth. The 2-local Hamiltonian problem encompasses NP. quant-ph/0301087, 2003. 2, 52, 53

[96] L.-A. Wu, P. Zanardi, and D. A. Lidar. Holonomic quantum computation in decoherence-free subspaces. *Physical Review Letters*, 95:130501, Sep 2005. DOI: 10.1103/PhysRevLett.95.130501. 30

[97] Mihalis Yannakakis. Computing the minimum fill-in is np-complete. *SIAM Journal on Algebraic and Discrete Methods*, 2(1):77–79, 1981. DOI: 10.1137/0602010. 65

[98] Paolo Zanardi and Mario Rasetti. Holonomic quantum computation. *Physics Letters A*, 264(2–3):94 – 99, 1999. DOI: 10.1016/S0375-9601(99)00803-8. 29

[99] Chi Zhang, Zhaohui Wei, and Anargyros Papageorgiou. Adiabatic quantum counting by geometric phase estimation. *Quantum Information Processing*, 9:369–383, 2010. 10.1007/s11128-009-0132-y. DOI: 10.1007/s11128-009-0132-y. 29

[100] M. Zimand. Weighted np optimization problems: Logical definability and approximation properties. *SIAM Journal on Computing*, 28(1):36–56, 1998. DOI: 10.1137/S0097539795285102. 63

Authors' Biographies

WILLIAM CRUZ-SANTOS

William Cruz-Santos is a full-time professor of Mathematics and Computer Science at the Computer Engineering at the Universidad Autónoma del Estado de México. Dr. Cruz-Santos's research interests include design of adiabatic quantum algorithms for solving NP-hard problems and simulation of quantum systems, as well as computational complexity analysis and algorithm design of classical algorithms. Dr. Cruz-Santos is particularly interested in the development of adiabatic quantum algorithms applied to computer vision problems from a combinatorial optimization point of view.

Dr. Cruz-Santos holds a B.Sc. in Computer Science from the Universidad Juárez Autónoma de Tabasco, as well as M.Sc. and Ph.D. degrees in Computer Science, both degrees from the Centro de Investigación y de Estudios Avanzados del IPN (Cinvestav-IPN). More information on Dr. Cruz-Santos's interests can be found on his web page `https://computacio n.cs.cinvestav.mx/~cwilliam/`.

GUILLERMO MORALES-LUNA

Guillermo Morales-Luna received the BSc degree in Mathematics from the Mexican National Polytechnic Institute in 1977, the MSc degree in Mathematics from Mexican Cinvestav-IPN, in 1978, and the PhD degree from the Mathematics Institute of the Polish Academy of Sciences in 1984. Since 1985 he is a researcher at Cinvestav-IPN. His research interests include cryptography, complexity theory, and mathematical logic. He is a Mexican national and he also holds Polish citizenship. He can be contacted at `gmorales@cs.cinvestav.mx` as well as at `http://delta. cs.cinvestav.mx/~gmorales`.

Printed in the United States
by Baker & Taylor Publisher Services